DC PANDEY'S

Physics QUICK BOOK

Last Minute Prep for

JEE, NEET, Class 11/12

DC PANDEY'S

Physics QUICK BOOK

Last Minute Prep for
JEE, NEET, Class 11/12

arihant
ARIHANT PRAKASHAN (Series), MEERUT

arihant
ARIHANT PRAKASHAN (Series), MEERUT

卐 Administrative & Production Offices

Regd. Office

'Ramchhaya' 4577/15, Agarwal Road, Darya Ganj, New Delhi -110002
Tele: 011- 47630600, 43518550

卐 Head Office

Kalindi, TP Nagar, Meerut (UP) - 250002, Tel: 0121-7156203, 7156204

卐 Sales & Support Offices

Agra, Ahmedabad, Bengaluru, Bareilly, Chennai, Delhi, Guwahati, Hyderabad, Jaipur, Jhansi, Kolkata, Lucknow, Nagpur & Pune.

卐 ISBN 978-93-25294-06-6

PO No : TXT-XX-XXXXXXX-X-XX

Published by Arihant Publications (India) Ltd.

For further information about the books published by Arihant, log on to www.arihantbooks.com or e-mail at info@arihantbooks.com

Follow us on

PREFACE

Dear students, it gives me immense pleasure to present this Physics Quick book for daily revision of concepts and formulae. For a long time I was planning to write this book but due to my extremely busy schedule I could not get time. But this lockdown period was an opportunity for me to completed this book.

I have seen that students want to make short notes for quick revision but due to lack of experience, they can't make the short notes properly. Sometimes they miss the important concepts/formulae or they note down those things which are not very useful. This book is an effort from my side on behalf of such students. I am sure, it will definitely solve your purpose. Try to revise at least one chapter per day till the final exam.

I am extremely thankful to Mr. Anoop Dhyani for their special contribution in this book.

Please feel free to share your suggestions and mistakes (if any in the book) to improve its revised editions from next year.

mail id. arihantcorrections@gmail.com

Thanks

DC Pandey

CONTENTS

CHAPTER 01

General Physics

Rules for Counting Significant Figures

***Rule* 1** All non-zero digits are significant. For example, 126.28 has five significant figures.

***Rule* 2** The zeros appearing between two non-zero digits are significant. For example, 6.025 has four significant figures.

***Rule* 3** Trailing zeros after decimal places are significant. Measurement $l = 6.400$ cm has four significant figures.

Let us take an example in its support.

Measurement	Accuracy	l lies between (in cm)	Significant figures	Remarks
$l = 6.4$ cm	0.1 cm	6.3-6.5	Two	
$l = 6.40$ cm	0.01 cm	6.39-6.41	Three	closer
$l = 6.400$ cm	0.001 cm	6.399-6.401	Four	more closer

Thus, the significant figures depend on the accuracy of measurement. More the number of significant figures, more accurate is the measurement.

***Rule* 4** The powers of ten are not counted as significant figures.

For example, 1.4×10^{-7} has only two significant figures 1 and 4.

***Rule* 5** If a measurement is less than one, then all zeros occurring to the left of last non-zero digit are not significant. For example, 0.0042 has two significant figures 4 and 2.

***Rule* 6** Change in units of measurement of a quantity does not change the number of significant figures. Suppose a measurement was done using mm scale and we get $l = 72$ mm (two significant figures).

We can write this measurement in other units also (without changing the number of significant figures) :

7.2 cm $\rightarrow$ Two significant figures

0.072 m $\rightarrow$ Two significant figures

0.000072 km $\rightarrow$ Two significant figures

7.2×10^7 nm $\rightarrow$ Two significant figures

***Rule* 7** The terminal or trailing zeros in a number without a decimal point are not significant. This also sometimes arises due to change of unit.

For example, 264 m = 26400 cm = 264000 mm

All have only three significant figures 2, 6 and 4.

Zeros at the end of a number are significant only, if they are behind a decimal point as in Rule 3. Otherwise, it is impossible to tell if they are significant.

For example, in the number 8200, it is not clear, if the zeros are significant or not. The number of significant digits in 8200 is at least two, but could be three or four.

To avoid uncertainty, use scientific notation to place significant zeros behind a decimal point

8.200×10^3 has four significant digits

8.20×10^3 has three significant digits

8.2×10^3 has two significant digits

Therefore, if it is not expressed in scientific notations, then write least number of significant digits. Hence, in the number 8200, take significant digits as two.

***Rule* 8** Exact measurements have infinite number of significant figures. For example,

10 bananas in a basket

46 students in a class

Speed of light in vacuum = 299,792,458 m/s (exact)

$$\pi = \frac{22}{7} \text{ (exact)}$$

All these measurements have infinite number of significant figures.

Rounding off a Digit

Following are the rules for rounding off a measurement

***Rule* 1** If the number lying to the right of cut off digit is less than 5, then the cut off digit is retained as such. However, if it is more than 5, then the cut off digit is increased by 1.

For example, $x = 6.24$ is rounded off to 6.2 to two significant digits and $x = 5.328$ is rounded off to 5.33 to three significant digits.

***Rule* 2** If the digit to be dropped is 5 followed by digits other than zero, then the preceding digit is increased by 1.

For example, $x = 14.252$ is rounded off to $x = 14.3$ to three significant digits.

***Rule* 3** If the digit to be dropped is simply 5 or 5 followed by zeros, then the preceding digit is left unchanged if it is even.

For example, $x = 6.250$ or $x = 6.25$ becomes $x = 6.2$ after rounding off to two significant digits.

***Rule* 4** If the digit to be dropped is 5 or 5 followed by zeros, then the preceding digit is raised by one if it is odd.

For example, $x = 6.350$ or $x = 6.35$ becomes $x = 6.4$ after rounding off to two significant digits.

Algebraic Operations with Significant Figures

Addition or Subtraction

Suppose in the measured values to be added or subtracted, the least number of digits after the decimal is n. Then, in the sum or difference, the number of significant digits after the decimal should also be n.

For example $1.2 + 3.45 + 6.789 = 11.439 \approx 11.4$

Here, the least number of significant digits after the decimal is one. Hence, the result will be 11.4 (when rounded off to smallest number of decimal places).

Multiplication or Division

Suppose in the measured values to be multiplied or divided, the least number of significant digits be n. Then, in the product or quotient, the number of significant digits should also be n.

For example $1.2 \times 36.72 = 44.064 \approx 44$

The least number of significant digits in the measured values are two. Hence, the result when rounded off to two significant digits will be 44. Therefore, the answer is 44.

Error Analysis

- **Least count**

Instrument	Its least count
mm scale	1 mm
Vernier callipers	0.1 mm
Screw gauge	0.01 mm
Stop watch	0.1 s
Temperature thermometer	1°C

- **True value**

  Usually the mean value a_m is taken as the true value. So,

$$a_m = \frac{a_1 + a_2 + \ldots + a_n}{n}$$

- **Absolute error**

$$\Delta a_1 = a_m - a_1$$
$$\Delta a_2 = a_m - a_2$$
$$\ldots \quad \ldots \quad \ldots$$
$$\Delta a_n = a_m - a_n$$

  Absolute error may be positive or negative.

- **Mean absolute error**

$$\Delta a_{\text{mean}} = \frac{|\Delta a_1| + |\Delta a_2| + \ldots + |\Delta a_n|}{n}$$

  The final result of measurement can be written as, $a = a_m \pm \Delta a_{\text{mean}}$.

- **Relative (or fractional) and percentage error**

$$\text{Relative error} = \frac{\Delta a_{\text{mean}}}{a_m}$$

$$\text{Percentage error} = \frac{\Delta a_{\text{mean}}}{a_m} \times 100$$

- **Error in sum or difference**

Let $x = a \pm b$

Then, $\Delta x = \pm (\Delta a + \Delta b)$

- **Error in product**

Let $x = ab$

Then, $\frac{\Delta x}{x} = \pm \left(\frac{\Delta a}{a} + \frac{\Delta b}{b} \right)$

- **Error in division**

Let $x = \frac{a}{b}$

Then, $\frac{\Delta x}{x} = \pm \left(\frac{\Delta a}{a} + \frac{\Delta b}{b} \right)$

- **Error in quantity raised to some power**

Let $x = \frac{a^n}{b^m}$

Then, $\frac{\Delta x}{x} = \pm \left(n \frac{\Delta a}{a} + m \frac{\Delta b}{b} \right)$

Experiments

1. Vernier Callipers

(i) $\text{VC} = \text{LC} = \frac{1\text{ MSD}}{n} = \frac{\text{smallest division on main scale}}{\text{number of divisions on vernier scale}} = 1\text{ MSD} - 1\text{ VSD}$

(ii) In ordinary vernier callipers, 1 MSD = 1 mm and $n = 10$

$\therefore$ $\text{VC or LC} = \frac{1}{10}\text{ mm} = 0.01\text{ cm}$

(iii) Total reading = $(N + n \times \text{VC})$ (N = main scale reading)

(iv) Zero correction = – Zero error

(v) Zero error is algebraically subtracted while the zero correction is algebraically added.

(vi) If zero of vernier scale lies to the right of zero of main scale, the error is positive. The actual length in this case is less than observed length.

(vii) If zero of vernier scale lies to the left of zero of main scale, the error is negative and the actual length is more than the observed length.

(viii) Positive zero error = $(N + x \times \text{VC})$

(ix) In negative zero error, suppose 8th vernier scale division coincides with the main scale division, then

Negative zero error = – [0.00 + 8 × VC] = – [0.00 + 8 × 0.01cm] = – 0.08 cm

2. Screw Gauge

$$\text{Least count} = \frac{\text{pitch}}{\text{number of divisions on circular scale}}$$

Total reading = $N + n \times \text{LC}$

If the zero of the circular scale advances beyond the reference line, the zero error is negative and zero correction is positive. If it is left behind the reference line, the zero error is positive and zero correction is negative.

For example, if zero of circular scale advances beyond the reference line by 5 divisions, zero correction = + 5 × (LC) and if the zero of circular scale is left behind the reference line by 5 divisions, zero correction = – 5 × (LC).

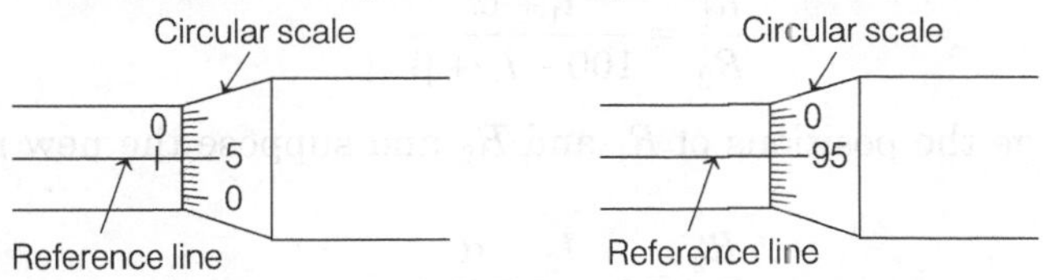

(a) Positive zero error (b) Negative zero error

Note *In negative zero error, 95th divisions of the circular scale is coinciding with the reference line. Hence, there are 5 divisions between zero mark on the circular scale and the reference line.*

3. Speed of Sound using Resonance Tube

(i) Result is independent of end correction

(ii) $v = 2f(l_2 - l_1)$, where f = frequency of tuning fork,
l_1 = first resonance length and l_2 = second resonance length.

(iii) End correction, $e = \dfrac{l_2 - 3l_1}{2}$

4. Meter Bridge Experiment

Meter bridge experiment is based on the principle of Wheatstone's bridge.

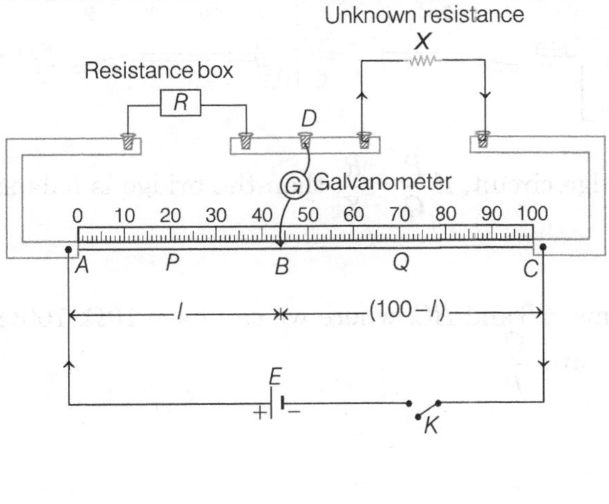

When current through galvanometer is zero or bridge is balanced, then

$$\frac{P}{Q}=\frac{R}{X}$$

$$\therefore \quad X = R\left(\frac{Q}{P}\right)=\left(\frac{100-l}{l}\right)R$$

End Corrections

In meter bridge, some extra length (under the metallic strips) comes at points A and C. Therefore, some additional length (α and β) should be included at the ends.
Here, α and β are called the end corrections. Hence, in place of l, we use $l+\alpha$ and in place of $100-l$, we use $100-l+\beta$.
To find α and β, use known resistors R_1 and R_2 in place of R and X and suppose we get null point length equal to l_1. Then,

$$\frac{R_1}{R_2}=\frac{l_1+\alpha}{100-l_1+\beta} \quad \text{...(i)}$$

Now, we interchange the positions of R_1 and R_2 and suppose the new null point length is l_2. Then,

$$\frac{R_2}{R_1}=\frac{l_2+\alpha}{100-l_2+\beta} \quad \text{...(ii)}$$

Solving Eqs. (i) and (ii), we can find α and β.

5. Post Office Box

Post office box also works on the principle of Wheatstone's bridge.

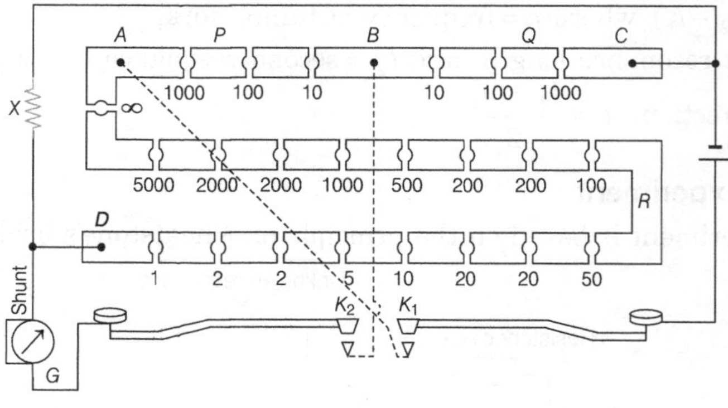

In a Wheatstone's bridge circuit, if $\frac{P}{Q}=\frac{R}{X}$, then the bridge is balanced. So, unknown resistance $X=\frac{Q}{P}R$.

P and Q are set in arms AB and BC, where we can have 10Ω, 100Ω or 1000 Ω resistances to set any ratio $\frac{Q}{P}$.

These arms are called ratio arms, initially we take $Q = 10\,\Omega$ and $P = 10\,\Omega$ to set $\frac{Q}{P} = 1$.

The unknown resistance (X) is connected between C and D and battery is connected across A and C.

Now, put resistance in part A to D such that the bridge gets balanced.

For this, keep on increasing the resistance with $1\,\Omega$ interval, check the deflection in galvanometer by first pressing key K_1, then galvanometer key K_2.

Suppose at $R = 4\Omega$, we get deflection towards left and at $R = 5\Omega$, we get deflection towards right.

Then, we can say that for balanced condition, R should lies between $4\,\Omega$ to $5\,\Omega$.

Now,
$$X = \frac{Q}{P} R = \frac{10}{10} R$$
$$= R = 4\,\Omega \text{ to } 5\,\Omega$$

To get closer value of X, in the second observation, let us choose $\frac{Q}{P} = \frac{1}{10}$, i.e. $\left(\frac{P = 100}{Q = 10}\right)$

Suppose now at $R = 42\Omega$, we get deflection towards left and at $R = 43\Omega$, deflection is towards right.

So, $R \in (42, 43)$.

Now, $X = \frac{Q}{P} R = \frac{10}{100} R = \frac{1}{10} R$, where $R \in (42, 43\,\Omega)$.

Now, to get further closer value, take $\frac{Q}{P} = \frac{1}{100}$ and so on.

The observation table is shown below

Resistance in the ratio arms		Resistance in arm AD (R) (ohm)	Direction of deflection	Unknown resistance $X = \frac{Q}{P} \times R$ (ohm)
AB (P) (ohm)	BC (Q) (ohm)			
10	10	4	Left	4 to 5
		5	Right	
100	10	40	Left (large)	(4.2 to 4.3)
		50	Right (large)	
		42	Left	
		43	Right	
1000	10	420	Left	4.25
		424	Left	
		425	No deflection	
		426	Right	

So, the correct value of X is $4.25\,\Omega$.

6. Focal Length of Concave Mirror

(i) $\frac{1}{v}$ *versus* $\frac{1}{u}$ **graph**

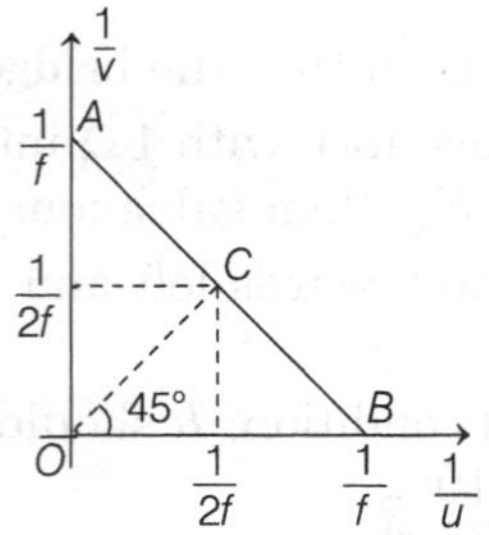

The coordinates of point C are $\left(\frac{1}{2f}, \frac{1}{2f}\right)$. The focal length of the concave mirror can be calculated by measuring the coordinates of either of the points A, B or C.

(ii) v *versus* u **graph**

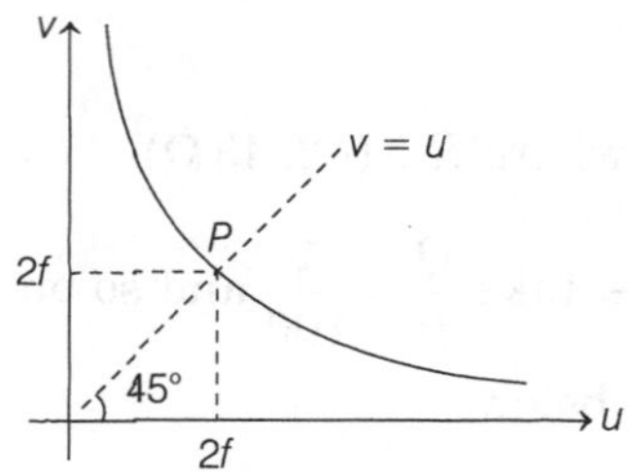

From u-v data, plot v *versus* u curve and draw a line bisecting the axis. Find the intersection point and equate them to $(2f, 2f)$.

(iii) **By joining v_n and u_n**

All lines intersect at a common point (f, f).

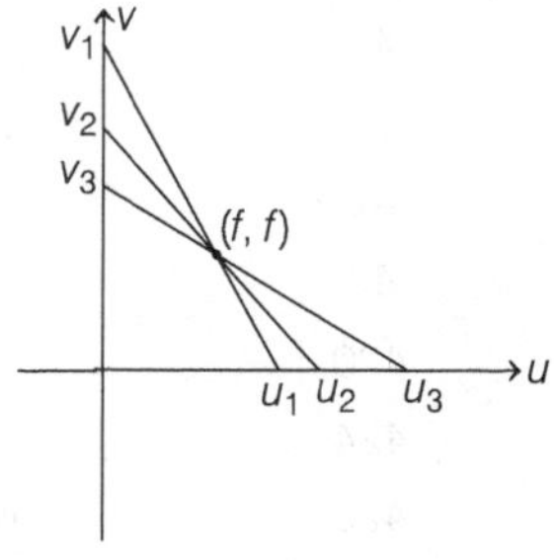

Find common intersection point and equate it to (f, f).

7. Focal Length of Convex Lens

(i) $\frac{1}{v}$ ***versus*** $\frac{1}{u}$ **graph**

The focal length of convex lens can be calculated by measuring the coordinates of either of the points *A, B* or *C*.

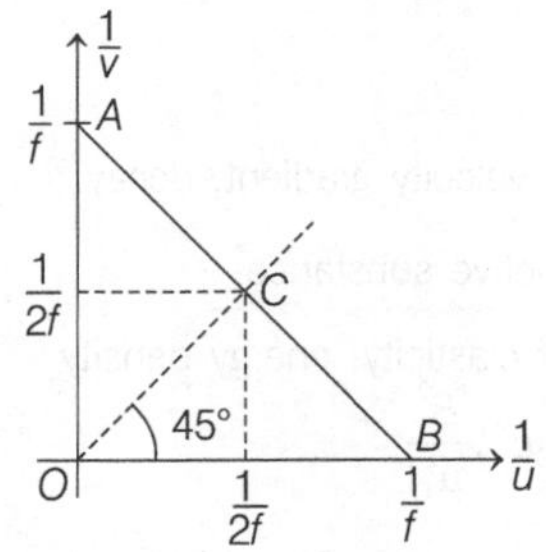

(ii) ***v versus u*** **graph**

By measuring the coordinates of point *C*, whose coordinates are $(2f, 2f)$, we can calculate the focal length of the lens.

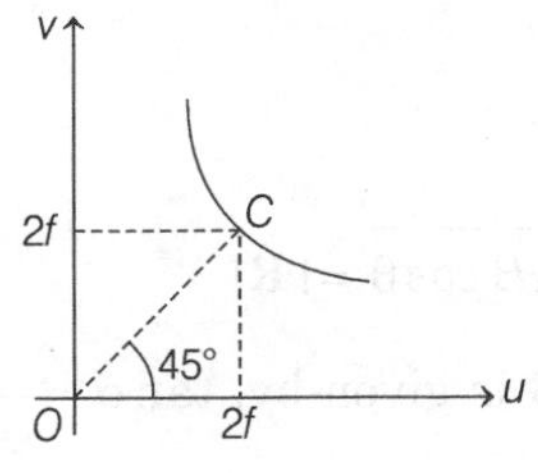

(iii) **By joining v_n and u_n**

All lines intersect at a common point $(-f, f)$.

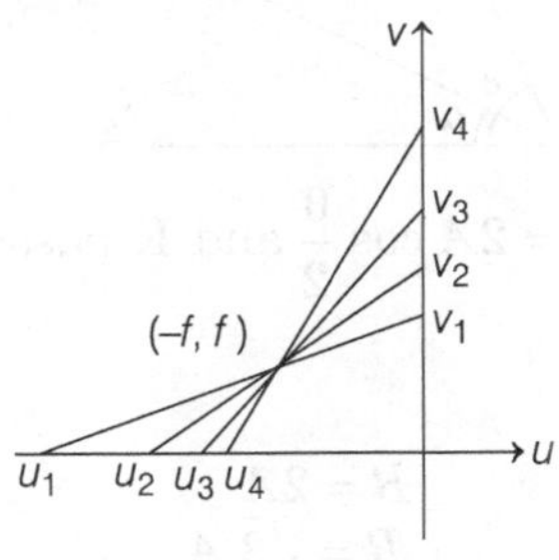

Find common intersection point and equate it to $(-f, f)$.

Note *All graphs are for real images.*

Physical quantities having the same dimensions

Physical Quantities or Combination of Physical Quantities	Dimensions
Angle, strain, $\sin\theta$, π, e^x	$[M^0 L^0 T^0]$
Work, energy, torque, Rhc	$[ML^2 T^{-2}]$
Time, $\frac{L}{R}$, CR, $\sqrt{LC}$	$[M^0 L^0 T]$
Frequency, ω, $\frac{R}{L}$, $\frac{1}{CR}$, $\frac{1}{\sqrt{LC}}$, velocity gradient, decay constant, activity of a radioactive substance	$[M^0 L^0 T^{-1}]$
Pressure, stress, modulus of elasticity, energy density (energy per unit volume), $\varepsilon_0 E^2$, $\frac{B^2}{\mu_0}$	$[ML^{-1} T^{-2}]$
Angular impulse, angular momentum, Planck's constant	$[ML^2 T^{-1}]$
Linear momentum, linear impulse	$[MLT^{-1}]$
Wavelength, radius of gyration, light year	$[M^0 LT^0]$
Velocity, $\frac{1}{\sqrt{\varepsilon_0\mu_0}}$, $\sqrt{\frac{GM}{R}}$, $\frac{E}{B}$	$[M^0 LT^{-1}]$

Vectors

- $R = |\mathbf{A} + \mathbf{B}| = \sqrt{A^2 + B^2 + 2AB\cos\theta} = |\mathbf{R}|$
- Angle of **R** from **A** towards **B** is given by, $\tan\alpha = \dfrac{B\sin\theta}{A + B\cos\theta}$

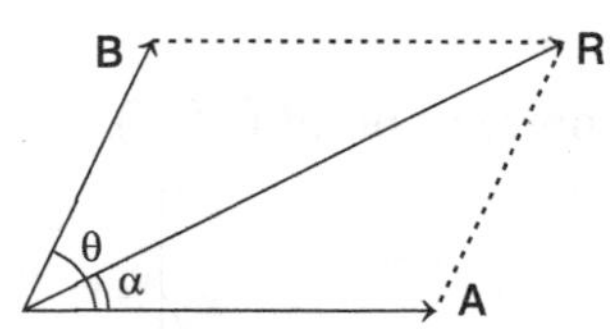

- If $|\mathbf{B}| = |\mathbf{A}| = A$(say), then $R = 2A\cos\frac{\theta}{2}$ and **R** passes along the bisector line of **A** and **B**.

  In this case, if

  $$\begin{aligned} &\theta = 0^\circ, && R = 2A \\ &\theta = 60^\circ, && R = \sqrt{3}A \\ &\theta = 90^\circ, && R = \sqrt{2}A \\ &\theta = 120^\circ, && R = A \\ \text{and} \quad &\theta = 180^\circ, && R = 0 \end{aligned}$$

- $S = |\mathbf{A} - \mathbf{B}| = \sqrt{A^2 + B^2 - 2AB\cos\theta} = |\mathbf{S}|$

  Here, θ is the angle between **A** and **B**, not the angle between **A** and $-\mathbf{B}$.

- Angle of **S** from **A** towards −**B** is given by, $\tan\alpha = \dfrac{B\sin\theta}{A - B\cos\theta}$

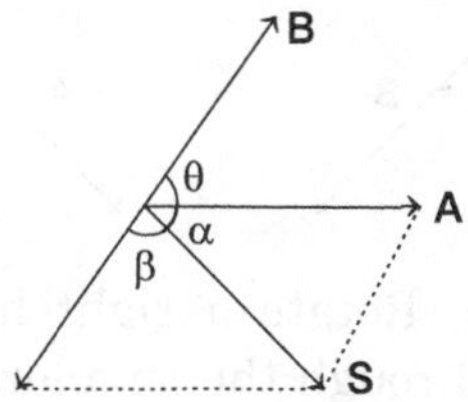

or angle of **S** from −**B** towards **A** is given by

$$\tan\beta = \frac{A\sin\theta}{B - A\cos\theta}$$

- If $|\mathbf{B}| = |\mathbf{A}| = A$ (say), then $S = 2A\sin\frac{\theta}{2}$ and **S** passes through the bisector line of **A** and −**B**.

In this case, if

	$\theta = 0°$,	$S = 0$
	$\theta = 60°$,	$S = A$
	$\theta = 90°$,	$S = \sqrt{2}\,A$
	$\theta = 120°$,	$S = \sqrt{3}\,A$
and	$\theta = 180°$,	$S = 2A$

- In the figure shown,

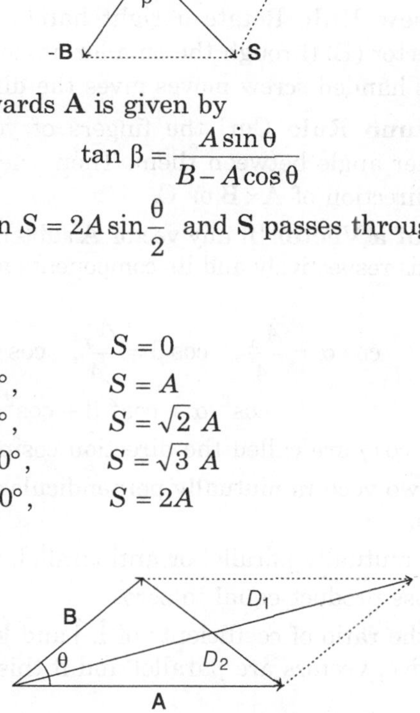

$$\text{diagonal, } D_1 = |\mathbf{A} + \mathbf{B} \text{ or } \mathbf{R}| = \sqrt{A^2 + B^2 + 2AB\cos\theta}$$

$$\text{diagonal, } D_2 = |\mathbf{A} - \mathbf{B} \text{ or } \mathbf{S}| = \sqrt{A^2 + B^2 - 2AB\cos\theta}$$

$$D_1 = D_2 = \sqrt{A^2 + B^2}, \text{ if } \theta = 90°$$

- $\mathbf{A}\cdot\mathbf{B} = AB\cos\theta$. Here, A and B are always positive as these are the magnitudes of **A** and **B**. Hence,

$0° \le \theta < 90°$, if $\mathbf{A}\cdot\mathbf{B}$ is positive

$90° < \theta \le 180°$, if $\mathbf{A}\cdot\mathbf{B}$ is negative.

and $\theta = 90°$, if $\mathbf{A}\cdot\mathbf{B}$ is zero.

- $|\mathbf{A}\times\mathbf{B}| = AB\sin\theta$
- $\mathbf{A}\times\mathbf{B} = -\mathbf{B}\times\mathbf{A}$
- **Direction of Vector Cross Product** When $\mathbf{C} = \mathbf{A}\times\mathbf{B}$, then the direction of **C** is at right angles to the plane containing the vectors **A** and **B**. The direction of **C** is determined by the right hand screw rule and right hand thumb rule.

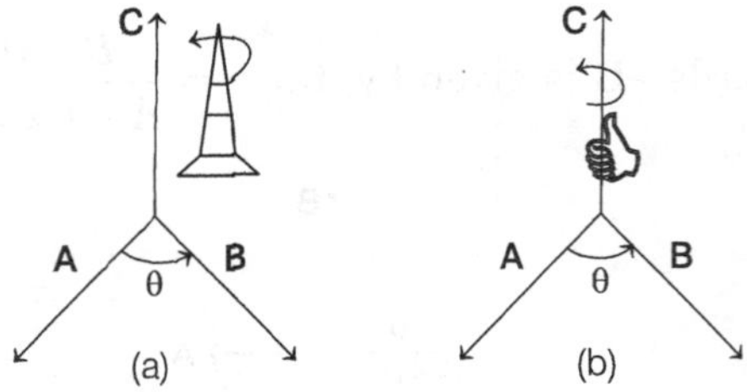

(a) **Right Hand Screw Rule** Rotate a right handed screw from first vector (**A**) towards second vector (**B**) through the smaller angle between them. The direction in which the right handed screw moves gives the direction of vector (**C**).

(b) **Right Hand Thumb Rule** Curl the fingers of your right hand from **A** to **B** through the smaller angle between them. Then, the direction of the erect thumb will point in the direction of $\mathbf{A} \times \mathbf{B}$ or **C**.

- **Direction Cosines of a Vector** If any vector **A** subtend angles α, β and γ with x-axis, y-axis and z-axis respectively and its components along these axes are A_x, A_y and A_z, then

$$\cos\alpha = \frac{A_x}{A}, \quad \cos\beta = \frac{A_y}{A}, \quad \cos\gamma = \frac{A_z}{A}$$

and $$\cos^2\alpha + \cos^2\beta + \cos^2\gamma = 1$$

Here, $\cos\alpha$, $\cos\beta$ and $\cos\gamma$ are called the direction cosines of **A** along x, y and z-axis.

- If we have to prove two vectors mutually perpendicular, then show their dot product equal to zero.
- To prove two vectors mutually parallel or antiparallel, we have two methods :

First Show their cross product equal to zero.

Second Show that the ratio of coefficients of $\hat{\mathbf{i}}$, $\hat{\mathbf{j}}$ and $\hat{\mathbf{k}}$ of two vectors is constant. If this constant is positive, vectors are parallel and if this constant is negative, vectors are antiparallel.

- **Angle between two vectors** In some cases, angle between two vectors can be obtained just by observation as given in following table :

A	B	θ between A and B
$2\hat{i}$	$6\hat{i}$	0°
$3\hat{j}$	$-5\hat{j}$	180°
$2\hat{i}$	$3\hat{j} - 4\hat{k}$	90°
$6\hat{i}$	$2\hat{i} + 2\hat{j}$	45°
$8\hat{i}$	$-4\hat{i} + 4\hat{j}$	135°

In general, angle between **A** and **B** can be obtained by the following relation,

$$\theta = \cos^{-1}\left(\frac{\mathbf{A}\cdot\mathbf{B}}{AB}\right)$$

It is not always, $\sin^{-1}\left\{\frac{|\mathbf{A}\times\mathbf{B}|}{AB}\right\}$

Let's Practice Explain the reason why θ is not always given by the following relation?

$$\theta = \sin^{-1}\left\{\frac{|\mathbf{A}\times\mathbf{B}|}{AB}\right\}$$

- Component of **A** along **B** $= A\cos\theta = \dfrac{\mathbf{A}\cdot\mathbf{B}}{B}$

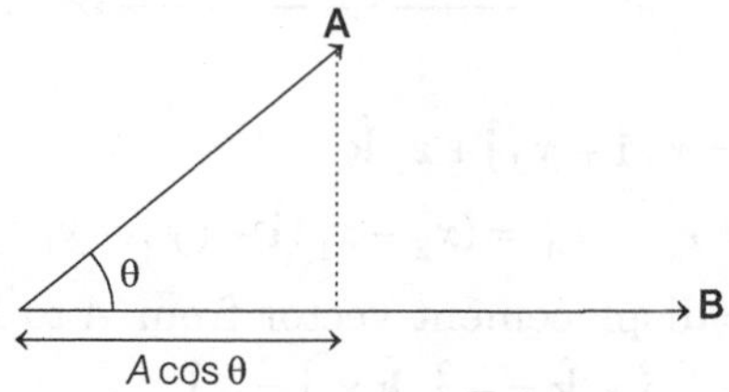

Similarly, component of **B** along **A** $= B\cos\theta = \dfrac{\mathbf{A}\cdot\mathbf{B}}{A}$

Component of **A** along **B** = component of **B** along **A**, if $|\mathbf{A}| = |\mathbf{B}|$ or $A = B$. Otherwise they are not equal.

- If resultant of n vectors is zero, of which $(n-1)$ vectors are known and only one vector is unknown, then this last unknown vector is equal and opposite to the resultant of $(n-1)$ known vectors.
- Vector sum of n vectors of same magnitudes is always zero if angle between two successive vectors is always $\left(\dfrac{360}{n}\right)^\circ$.
- If resultant of **A** and **B** is along PQ, then components of **A** and **B** perpendicular to PQ or along MN should be equal and opposite.

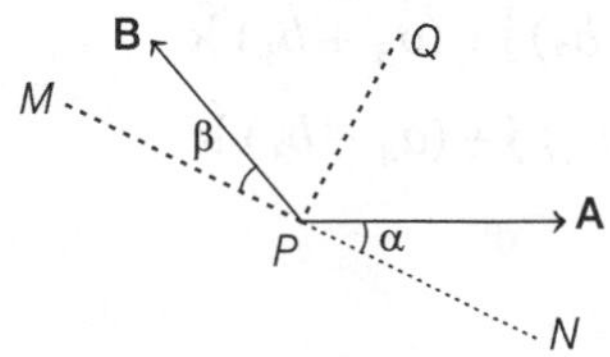

$\Rightarrow$ $$A\cos\alpha = B\cos\beta$$

and the resultant along PQ is, $$R = A\sin\alpha + B\sin\beta$$

- A unit vector perpendicular to both **A** and **B**

$$\hat{\mathbf{C}} = \pm\frac{\mathbf{A}\times\mathbf{B}}{|\mathbf{A}\times\mathbf{B}|}$$

- $\hat{\mathbf{i}}\cdot\hat{\mathbf{i}} = \hat{\mathbf{j}}\cdot\hat{\mathbf{j}} = \hat{\mathbf{k}}\cdot\hat{\mathbf{k}} = 1, \hat{\mathbf{i}}\cdot\hat{\mathbf{j}} = \hat{\mathbf{j}}\cdot\hat{\mathbf{k}} = \hat{\mathbf{i}}\cdot\hat{\mathbf{k}} = 0$
- If coordinates of point A are (x_1, y_1, z_1) and coordinates of point B are (x_2, y_2, z_2), then

$\mathbf{r}_A$ = position vector of $A = x_1\hat{\mathbf{i}} + y_1\hat{\mathbf{j}} + z_1\hat{\mathbf{k}}$

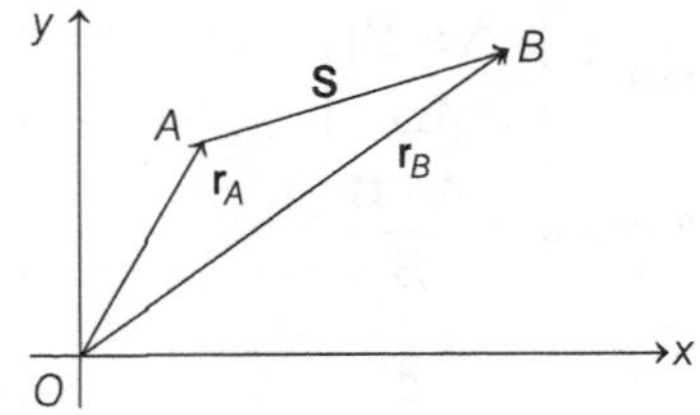

$$\mathbf{r}_B = x_2\,\hat{\mathbf{i}} + y_2\,\hat{\mathbf{j}} + z_2\,\hat{\mathbf{k}}$$

$$\mathbf{S} = \mathbf{r}_B - \mathbf{r}_A = (x_2 - x_1)\,\hat{\mathbf{i}} + (y_2 - y_1)\,\hat{\mathbf{j}} + (z_2 - z_1)\,\hat{\mathbf{k}}$$

= displacement vector from A to B

- $\hat{\mathbf{i}} \times \hat{\mathbf{j}} = \hat{\mathbf{k}}$, $\hat{\mathbf{j}} \times \hat{\mathbf{k}} = \hat{\mathbf{i}}$, $\hat{\mathbf{k}} \times \hat{\mathbf{i}} = \hat{\mathbf{j}}$, $\hat{\mathbf{i}} \times \hat{\mathbf{k}} = -\hat{\mathbf{j}}$, $\hat{\mathbf{k}} \times \hat{\mathbf{j}} = -\hat{\mathbf{i}}$

  $\hat{\mathbf{j}} \times \hat{\mathbf{i}} = -\hat{\mathbf{k}}$, $\hat{\mathbf{i}} \times \hat{\mathbf{i}} = \hat{\mathbf{j}} \times \hat{\mathbf{j}} = \hat{\mathbf{k}} \times \hat{\mathbf{k}} =$ a null vector

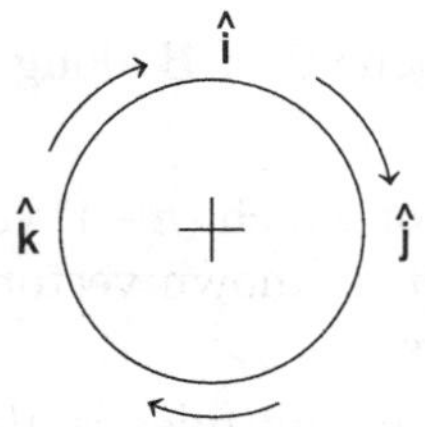

- If vectors are given in terms of $\hat{\mathbf{i}}$, $\hat{\mathbf{j}}$ and $\hat{\mathbf{k}}$,

  let $\mathbf{A} = a_1\hat{\mathbf{i}} + a_2\hat{\mathbf{j}} + a_3\hat{\mathbf{k}}$ and $\mathbf{B} = b_1\hat{\mathbf{i}} + b_2\hat{\mathbf{j}} + b_3\hat{\mathbf{k}}$, then

  (i) $|\mathbf{A}| = A = \sqrt{a_1^2 + a_2^2 + a_3^2}$ and $|\mathbf{B}| = B = \sqrt{b_1^2 + b_2^2 + b_3^2}$

  (ii) $\mathbf{A} + \mathbf{B} = (a_1 + b_1)\,\hat{\mathbf{i}} + (a_2 + b_2)\,\hat{\mathbf{j}} + (a_3 + b_3)\,\hat{\mathbf{k}}$

  (iii) $\mathbf{A} - \mathbf{B} = (a_1 - b_1)\,\hat{\mathbf{i}} + (a_2 - b_2)\,\hat{\mathbf{j}} + (a_3 - b_3)\,\hat{\mathbf{k}}$

  (iv) $\mathbf{A} \cdot \mathbf{B} = a_1 b_1 + a_2 b_2 + a_3 b_3$

  (v) $|\mathbf{A} \times \mathbf{B}| = \begin{vmatrix} \hat{\mathbf{i}} & \hat{\mathbf{j}} & \hat{\mathbf{k}} \\ a_1 & a_2 & a_3 \\ b_1 & b_2 & b_3 \end{vmatrix} = (a_2 b_3 - b_2 a_3)\,\hat{\mathbf{i}} + (b_1 a_3 - b_3 a_1)\,\hat{\mathbf{j}} + (a_1 b_2 - b_1 a_2)\,\hat{\mathbf{k}}$

  (vi) Component of **A** along **B**

  $$= A\cos\theta = \frac{\mathbf{A} \cdot \mathbf{B}}{B} = \frac{a_1 b_1 + a_2 b_2 + a_3 b_3}{\sqrt{b_1^2 + b_2^2 + b_3^2}}$$

  (vii) Unit vector parallel to **A**

  $$= \hat{\mathbf{A}} = \frac{\mathbf{A}}{A} = \frac{a_1\hat{\mathbf{i}} + a_2\hat{\mathbf{j}} + a_3\hat{\mathbf{k}}}{\sqrt{a_1^2 + a_2^2 + a_3^2}}$$

(viii) Angle between **A** and **B**,

$$\theta = \cos^{-1}\left(\frac{\mathbf{A}\cdot\mathbf{B}}{AB}\right)$$

$$\therefore \qquad \theta = \cos^{-1}\left(\frac{a_1 b_1 + a_2 b_2 + a_3 b_3}{\sqrt{a_1^2 + a_2^2 + a_3^2} \times \sqrt{b_1^2 + b_2^2 + b_3^2}}\right)$$

- (**A** + **B**) is perpendicular to (**A** − **B**), if $A = B$.
- (**A** × **B**) is perpendicular to both **A** and **B** separately, or it is perpendicular to the plane formed by **A** and **B**.
- $\hat{\mathbf{i}} \times \hat{\mathbf{j}}$ should always be in the direction of $\hat{\mathbf{k}}$.
- Pressure is a scalar quantity, not a vector quantity. It has magnitude but no direction sense associated with it. Pressure acts in all directions at a point inside a fluid.
- Surface tension is scalar quantity because it has no specific direction. Current is also a scalar quantity.
- Stress and moment of inertia are tensor quantities.
- To qualify as a vector, a physical quantity must not only possess magnitude and direction but must also satisfy the parallelogram law of vector addition.
  For example, the finite rotation of a rigid body about a given axis has magnitude (the angle of rotation) and also direction (the direction of the axis) but it is not a vector quantity.
  This is so far the simple reason that the two finite rotations of the body do not add up in accordance with the law of vector addition.
  However, if the rotation be small or infinitesimal, it may be regarded as a vector quantity.
- Area can behave either as a scalar or a vector and how it behaves depends on circumstances.
- Area (vector), dipole moment and current density are defined as vectors with specific direction.
- The area of triangle bounded by vectors **A** and **B** is $\frac{1}{2}|\mathbf{A}\times\mathbf{B}|$.

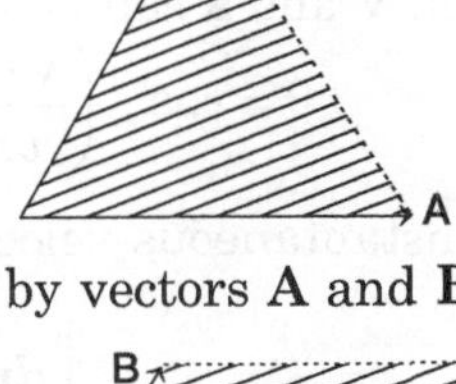

- Area of parallelogram bounded by vectors **A** and **B** is $|\mathbf{A}\times\mathbf{B}|$.

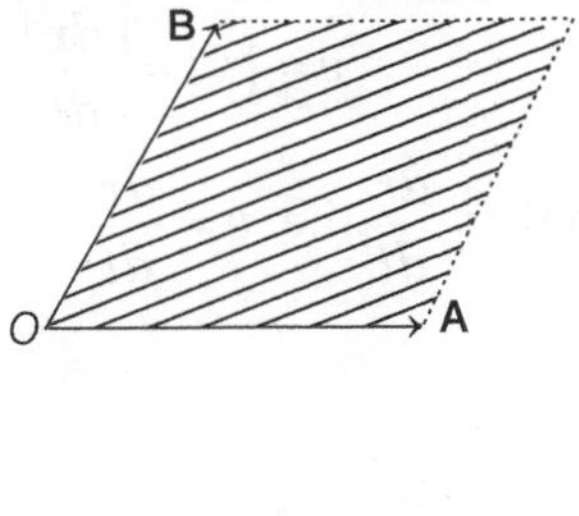

CHAPTER 02

Kinematics I

General Points

- If a particle is just dropped from a moving body, then just after dropping, velocity of the particle (not acceleration) is equal to the velocity of the moving body at that instant.
- If y (may be velocity, acceleration, etc.) is a function of time or $y = f(t)$ and we want to find the average value of y between a time interval of t_1 and t_2, then

$$< y >_{t_1 \text{ to } t_2} = \text{average value of } y \text{ between } t_1 \text{ and } t_2$$

$$= \frac{\int_{t_1}^{t_2} f(t)\, dt}{t_2 - t_1}$$

If $f(t)$ is a linear function of t, then $y_{av} = \dfrac{y_f + y_i}{2}$

Here, y_f = final value of y and y_i = initial value of y.

- Angle between velocity vector **v** and acceleration vector **a** decides whether the speed of particle is increasing, decreasing or constant.

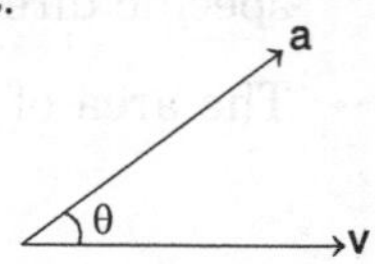

Speed increases, if $0° \le \theta < 90°$

Speed decreases, if $90° < \theta \le 180°$

Speed is constant, if $\theta = 90°$

The angle θ between **v** and **a** can be obtained by the following relation,

$$\theta = \cos^{-1}\left(\frac{\mathbf{v} \cdot \mathbf{a}}{va}\right)$$

- The magnitude of instantaneous velocity is called the instantaneous speed, i.e.

$$v = |\mathbf{v}| = \left|\frac{d\mathbf{r}}{dt}\right|$$

Speed is not equal to $\dfrac{dr}{dt}$, i.e. $v \ne \dfrac{dr}{dt}$

where, r is the modulus of radius vector $\mathbf{r}$ because in general $|d\mathbf{r}| \neq dr$.
For example, when $\mathbf{r}$ changes only in direction, i.e. if a point moves in a circle, then r = constant, $dr = 0$ but $|d\mathbf{r}| \neq 0$.

- The sign of acceleration does not tell us whether the particle's speed is increasing or decreasing. This sign of acceleration depends upon the choice of the positive direction of the axis.
  For example, if the vertically upward direction is chosen to be the positive direction, the acceleration due to gravity is negative.
  If a particle is falling under gravity, this acceleration though negative, results in increase in speed.
  For a particle thrown upward, the same negative acceleration (of gravity) results in decrease in speed.
- The zero velocity of a particle at any instant does not necessarily imply zero acceleration at that instant. A particle may be momentarily at rest and yet have non-zero acceleration.
  For example, a particle thrown up has zero velocity at its highest point but the acceleration at that instant continues to be the acceleration due to gravity.
- **Reaction time** When a situation demands our immediate action, it takes some time before we really respond. Reaction time is the time, a person takes to observe, think and act.
  For example, if a person is driving and suddenly a boy appears on the road, then the time elapsed before he applies the brakes of the car is the reaction time.

Basic Definitions

- Displacement $\mathbf{s} = \mathbf{r}_f - \mathbf{r}_i = (x_f - x_i)\,\hat{\mathbf{i}} + (y_f - y_i)\,\hat{\mathbf{j}} + (z_f - z_i)\,\hat{\mathbf{k}}$
- Distance = actual path length
- Average velocity $= \dfrac{\text{total displacement}}{\text{total time}} = \dfrac{s}{t}$
- Average speed $= \dfrac{\text{total distance}}{\text{total time}} = \dfrac{d}{t}$
- Average acceleration $= \dfrac{\text{change in velocity}}{\text{time}} = \dfrac{\Delta \mathbf{v}}{\Delta t}$
  $$= \frac{\mathbf{v}_f - \mathbf{v}_i}{t}$$
- Instantaneous velocity $= \dfrac{d\mathbf{s}}{dt}$ or $\dfrac{d\mathbf{r}}{dt}$
- Instantaneous acceleration = rate of change of velocity
  $$= \frac{d\mathbf{v}}{dt} = \frac{d^2\mathbf{s}}{dt^2} = \frac{d^2\mathbf{r}}{dt^2}$$

In one dimensional motion : (say along x-axis)

- Displacement $s = x_f - x_i$
- Instantaneous velocity $v = \frac{dx}{dt}$ or $\frac{ds}{dt}$ = slope of x-t or s-t graph
- Average velocity $= \frac{s}{t} = \frac{x_f - x_i}{t}$
- Instantaneous acceleration $a = \frac{dv}{dt}$ = slope of v-t graph
- Average acceleration $= \frac{\Delta v}{\Delta t} = \frac{v_f - v_i}{t}$

Note *In the above expressions, all vectors quantities are to be substituted with sign (in 1-D motion).*

Uniform Motion

In a uniform motion, velocity remains constant and acceleration is zero. Since, velocity is constant, so motion will be one dimensional without change in direction with same speed.

Hence, distance (d) and displacement (s) will be same in magnitude.

Due to direction, displacement may be positive or negative but distance is always positive. Equations are as follows

(i) $a = 0$

(ii) v = constant

(iii) $d = |\mathbf{s}| = |\mathbf{v}|\, t$ or, sometimes we simply write, $d = s = vt$

Note *Most of the problems of average speed are based on uniform motion.*

One Dimensional Motion with Uniform Acceleration

- Important equations for one dimensional motion with uniform acceleration are as folows

(i) $v = u + at$

(ii) $s = ut + \frac{1}{2}at^2$

(iii) $s' = s_0 + ut + \frac{1}{2}at^2$

(iv) $v^2 = u^2 + 2as$

(v) $s_t = (u + at) - \frac{a}{2}$

- While using above equations, take a sign convention and substitute all vector quantities (v, u, a, s and s_t) with sign.
- In equation $s = ut + \frac{1}{2}at^2$, s is the displacement measured from the starting point ($t = 0$).
- s_t is the displacement (not distance) in t^{th} second or between the time $(t-1)$ and t.
- In the above equations, s and s' both are displacements from time $t = 0$ to $t = t$. But s is measured from the point $t = 0$, while s' is measured from any other point (say P). Further, s_0 is the displacement of starting point ($t = 0$) from point P in that case.

- For small heights, if the motion is taking place under gravity, then acceleration is always constant (= acceleration due to gravity), i.e. $9.8\ \text{m/s}^2$ ($\approx 10\ \text{m/s}^2$) in downward direction. According to sign convention, upward direction is positive and downward direction is negative. Therefore, $a = g = -9.8\ \text{m/s}^2 \approx -10\ \text{m/s}^2$
- In most of the problems of time calculations, equation $s = ut + \frac{1}{2}at^2$ is useful. But s has to be measured from the starting point ($t = 0$).

  (i) If a particle is projected upwards with velocity u, then

  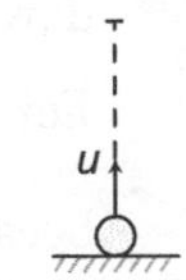

  (a) maximum height attained by the particle, $h = \frac{u^2}{2g}$

  (b) time of ascent = time of descent $= \frac{u}{g} \Rightarrow$ Total time of flight $= \frac{2u}{g}$

  (ii) If a particle is released from rest from a height h (also called free fall), then

  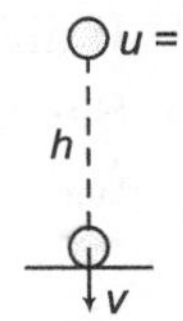

  (a) velocity of particle at the time of striking with ground, $v = \sqrt{2gh}$

  (b) time of descent (also called free fall time), $t = \sqrt{\frac{2h}{g}}$

Note *In the above results, air resistance has been neglected and we have already substituted the signs of u, g, etc. So, you have to substitute only their magnitudes.*

- A particle is thrown upwards with velocity u. Suppose it takes time t to reach its highest point, then distance travelled in last second is independent of u.

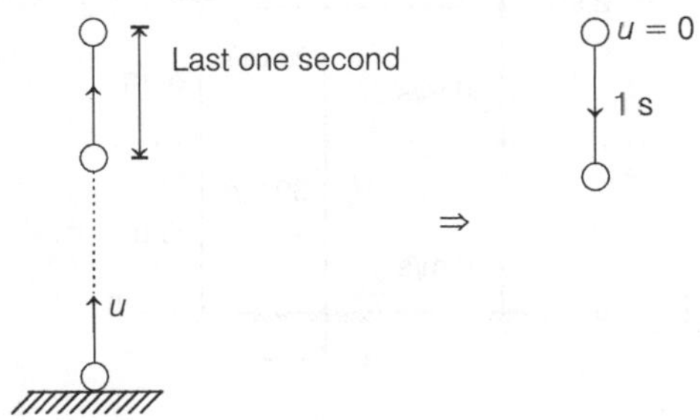

This is because this distance is equal to the distance travelled in first second of a freely falling object. Thus,

$$s = \frac{1}{2}g \times (1)^2 = \frac{1}{2} \times 10 \times 1 = 5\ \text{m}$$

- If a particle is just dropped ($u = 0$) under gravity, then after time t, velocity of particle in downward direction will be gt or $10t$ (if $g = 10\ \text{m/s}^2$) and total distance fallen downward would be $\frac{1}{2}gt^2$ or $5t^2$.

  Now in 1 s, it will fall 5 m, in 2 s, it will fall 20 m, in 3 s, it will fall 45 m and in 4 s, it will fall 80 m and so on.

  – In Ist second, it will fall 5 m.

  – In 2nd second it will fall $(20 - 5)\ \text{m} = 15\ \text{m}$

– In 3rd second, it will fall $(45-20)\,m = 25\,m$ and so on.

The ratio of these distances would be 5m : 15 m : 25 m or 1 : 3 : 5.
This result can be generalised with every constant acceleration problems for any equal time interval (not necessarily 1s), if $u = 0$.

$s_{0\to t} : s_{t\to 2t} : s_{2t\to 3t} = 1:3:5$ (if $u = 0$ and $a =$ constant)

$g = 10\ m/s^2$ (downwards) means in every second velocity changes by 10 m/s in downward direction.

For example, if a particle is projected upwards by 35 m/s, then $t_{up} = \frac{35}{10} = 3.5$ s because in every second, velocity is changing by 10 m/s in downward direction. At $t = 0$, $v = 35$ m/s (upwards), at $t = 1$ s, it will remain 25 m/s (upwards). So, it will take total 3.5 s to become zero.

If air resistance is neglected, then $t_{down} = 3.5$ s or $t_{total} = 7$ s

- **Second's diagram** Suppose a particle is projected upwards with 40 m/s, then $t_{up} = 4$ s and t_{down} is also equal to 4s.

$\therefore$ $t_{total} = 8$ s

Now, let us make a diagram showing the position and velocity at every second.

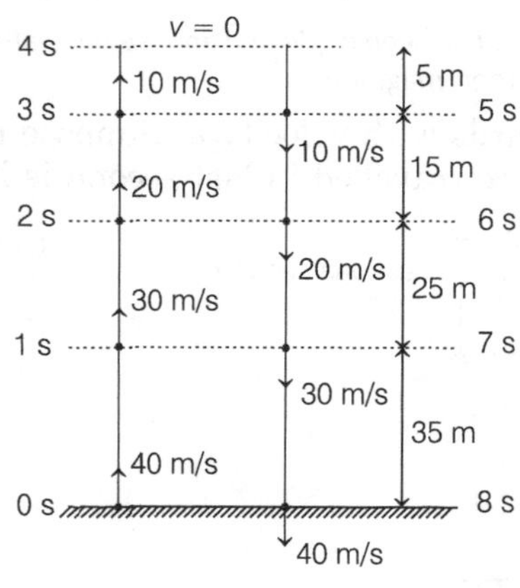

One Dimensional Motion with Non-uniform Acceleration

- s-$t \to v$-$t \to a$-$t \to$ Differentiation
- a-$t \to v$-$t \to s$-$t \to$ Integration
- **Equations of differentiation**

$$v = \frac{ds}{dt} \quad \text{and} \quad a = \frac{dv}{dt} = v \cdot \frac{dv}{ds}$$

- **Equations of integration**

$$\int ds = \int v dt, \int dv = \int a dt, \int v dv = \int a ds$$

In first integration equation, v should be either a constant or a function of t.
In second equation, a should be either a constant or a function of t. Similarly, in third equation, a should be either a constant or a function of s.

Two or Three Dimensional Motion with Uniform Acceleration

(i) $\mathbf{v} = \mathbf{u} + \mathbf{a}t$ (ii) $\mathbf{s} = \mathbf{u}t + \frac{1}{2}\mathbf{a}t^2$ (iii) $\mathbf{v} \cdot \mathbf{v} = \mathbf{u} \cdot \mathbf{u} + 2\mathbf{a} \cdot \mathbf{s}$

Two or Three Dimensional Motion with Non-uniform Acceleration

(i) $\mathbf{v} = \frac{d\mathbf{s}}{dt}$ or $\frac{d\mathbf{r}}{dt}$ (ii) $\mathbf{a} = \frac{d\mathbf{v}}{dt}$

(iii) $\int d\mathbf{v} = \int \mathbf{a} \cdot dt$ (iv) $\int d\mathbf{s} = \int \mathbf{v} \cdot dt$

Graphs

- Slope of s - t graph = velocity
  Slope of v - t graph = acceleration
  Area under v - t graph = displacement and
  Area under a - t graph = change in velocity.
- **Uniform motion** v = constant, $a = 0, s = vt$

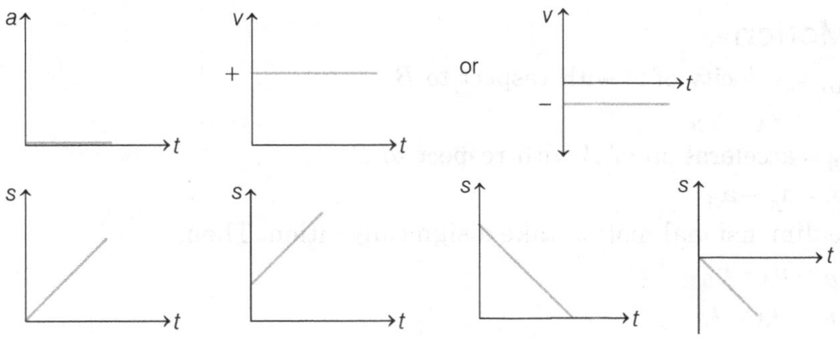

Since $a = 0$, therefore slope of v-t graph = 0
Further v = constant, therefore slope of s - t graph = constant.
- **Uniformly accelerated or retarded motion**

$$a = \text{constant},\ v = u \pm at,\ s = ut \pm \frac{1}{2}at^2$$

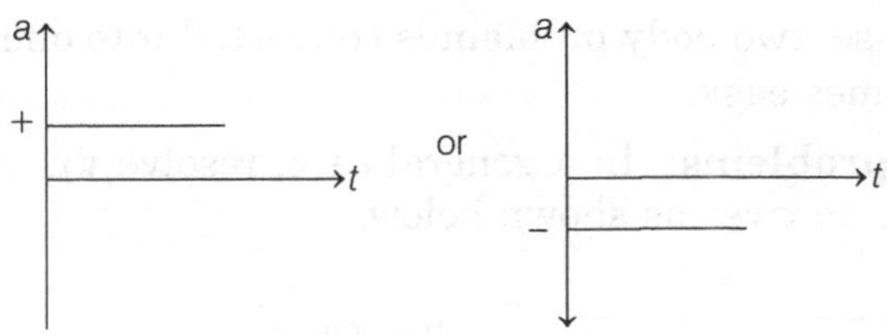

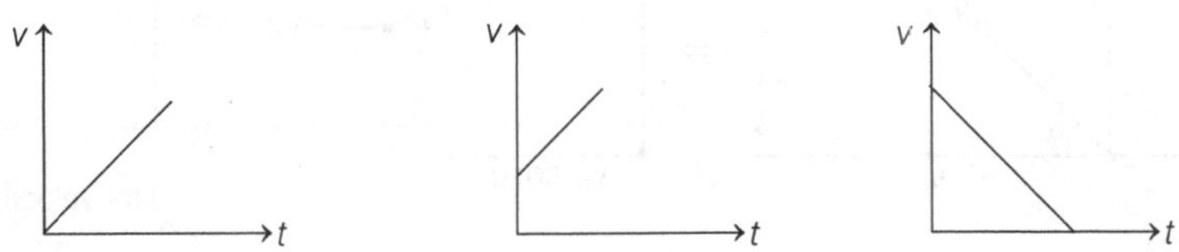

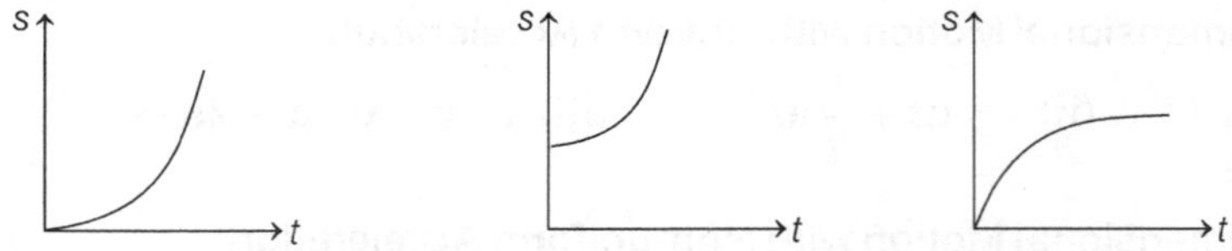

Since a = constant, therefore slope of v - t graph = constant.

Further, v is increasing or decreasing, therefore slope of s - t graph should either increase or decrease.

Note *From the given s-t graph, we can find sign of velocity and acceleration. For example, in the given graph, slope at t_1 and t_2 both are positive. Therefore, v_{t_1} and v_{t_2} are positive. Further, slope at t_2 > slope at t_1. Therefore, $v_{t_2} > v_{t_1}$. Hence, acceleration of the particle is also positive.*

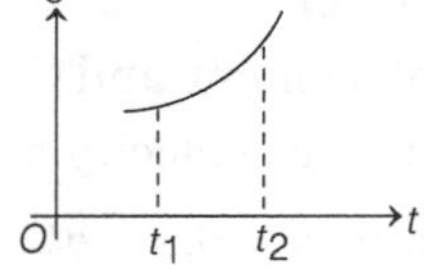

Relative Motion

- $\mathbf{v}_{AB}$ = velocity of A with respect to B
  $= \mathbf{v}_A - \mathbf{v}_B$
- $\mathbf{a}_{AB}$ = acceleration of A with respect to B
  $= \mathbf{a}_A - \mathbf{a}_B$

In one dimensional motion take a sign convention. Then,

- $v_{AB} = v_A - v_B$
- $a_{AB} = a_A - a_B$

Uses of Relative Motion

- **Minimum distance or collision problems** When two bodies are in motion, the questions like, the minimum distance between them or the time when one body overtakes the other can be solved easily by the principle of relative motion.

  In these types of problems, one body is assumed to be at rest and the relative motion of the other body is considered.

  By assuming so, two body problem is converted into one body problem and the solution becomes easy.

- **River boat problems** In a general case, resolve $\mathbf{v}_{br}$ along the river and perpendicular to river as shown below.

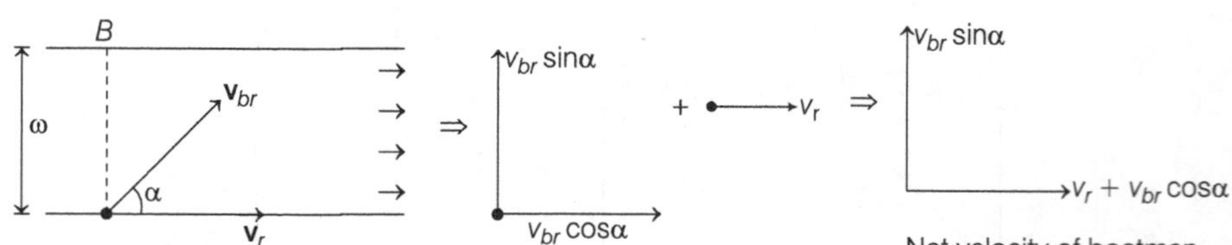

Net velocity of boatman

Now, the boatman will cross the river with component of $\mathbf{v}_{br}$ perpendicular to river (= v_{br} sin α in above case)

$$\therefore \quad t = \frac{\omega}{v_{br} \sin \alpha}$$

To cross the river in minimum time, why to take help of component of $\mathbf{v}_{br}$ (which is always less than v_{br}), the complete vector $\mathbf{v}_{br}$ should be kept perpendicular to the river current. Due to the other component $v_r + v_{br} \cos \alpha$, boatman will drift along the river by a distance $x = (v_r + v_{br} \cos \alpha)$ (time)

To reach a point *B*, which is just opposite to the starting point *A*, net velocity of boatman $\mathbf{v}_b$ or the vector sum of $\mathbf{v}_r$ and $\mathbf{v}_{br}$ should be along *AB*.

The velocity diagram is as under

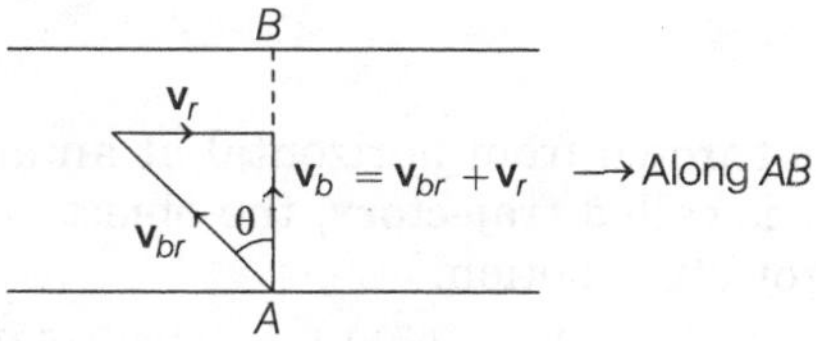

From the diagram, we can see that,

$$|\mathbf{v}_b| \text{ or } v_b = \sqrt{v_{br}^2 - v_r^2} \quad \text{...(i)}$$

$$\text{Time,} \quad t = \frac{\omega}{v_b} = \frac{\omega}{\sqrt{v_{br}^2 - v_r^2}}$$

$$\text{Drift } x = 0 \text{ and } \sin \theta = \frac{v_r}{v_{br}} \text{ or } \theta = \sin^{-1}\left(\frac{v_r}{v_{br}}\right)$$

From Eq. (i), we can see that this case is possible, if $v_{br} > v_r$ otherwise, v_b is either zero or imaginary.

If the boatman rows his boat along the river (downstream), then net velocity of boatman will be $v_{br} + v_r$.

If he rows along the river upstream, then net velocity of boatman will be $v_{br} \sim v_r$.

Note *Aircraft wind problems can also be solved in the similar manner. Velocity of boatman w.r.t. river ($\mathbf{v}_{br}$) can be replaced by velocity of aircraft w.r.t. wind ($\mathbf{v}_{aw}$). Velocity of river ($\mathbf{v}_r$) can be replaced by velocity of wind ($\mathbf{v}_w$) and net velocity of boatman ($\mathbf{v}_b$) can be replaced by net velocity of aircraft ($\mathbf{v}_a$).*

- **Rain umbrella problems** A person should hold his umbrella in the direction of $\mathbf{v}_{rp}$ or $\mathbf{v}_r - \mathbf{v}_p$ or velocity of rain w.r.t. person.

CHAPTER 03

Kinematics II

Projectile Motion

- When any object is thrown from horizontal at an angle θ except 90°, then the path followed by it is called trajectory, the object is called projectile and its motion is called projectile motion.
- Every projectile motion can be solved by either of the following two methods:

  ***Method* 1** Projectile motion is a two dimensional motion with constant acceleration. Therefore, we can use the equations

$$\mathbf{v} = \mathbf{u} + \mathbf{a}t \text{ and } \mathbf{s} = \mathbf{u}t + \frac{1}{2}\mathbf{a}t^2$$

For example, in the shown figure

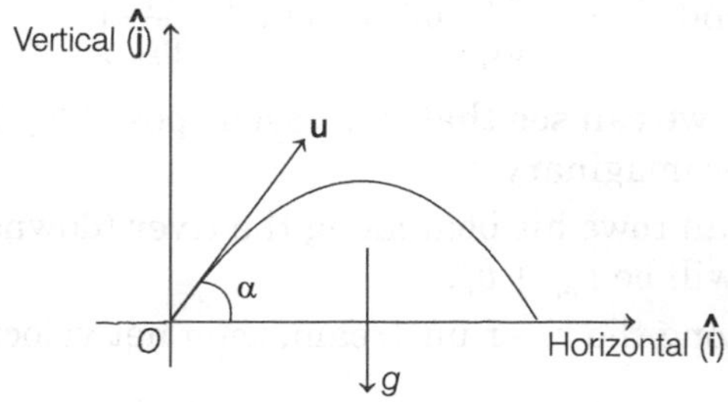

$$\mathbf{u} = u \cos \alpha \hat{\mathbf{i}} + u \sin \alpha \hat{\mathbf{j}} \quad \text{and} \quad \mathbf{a} = -g\hat{\mathbf{j}}$$

Note *In all problems, value of* **a** *(=* **g***) will be same, only* **u** *will be different.*

***Method* 2** In this method, select two mutually perpendicular directions x and y and find the two components of initial velocity and acceleration along these two directions, i.e. find u_x, u_y, a_x and a_y.

Now, apply the appropriate equation (s) of the following six equations :

$$\left.\begin{aligned} v_x &= u_x + a_x t \\ s_x &= u_x t + \frac{1}{2} a_x t^2 \\ v_x^2 &= u_x^2 + 2a_x s_x \end{aligned}\right] \rightarrow \text{Along } x\text{-axis}$$

$$\left.\begin{aligned} v_y &= u_y + a_y t \\ s_y &= u_y t + \frac{1}{2} a_y t^2 \\ v_y^2 &= u_y^2 + 2a_y s_y \end{aligned}\right] \rightarrow \text{Along } y\text{-axis}$$

Substitute v_x, u_x, a_x, s_x, v_y , u_y, a_y and s_y with proper signs but choosing one direction as positive and other as the negative along both axes.

In most of the problems, $s = ut + \frac{1}{2} at^2$ equation is useful for time calculation.

Under normal projectile motion, *x*-axis is taken along horizontal direction and *y*-axis along vertical direction. In projectile motion along an inclined plane, *x*-axis is normally taken along the plane and *y*-axis perpendicular to the plane.

Two simple cases are shown below.

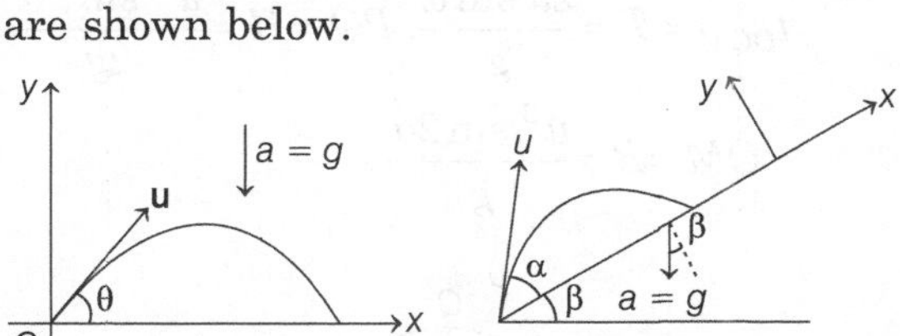

(a) (b)

In Fig. (a),

$$u_x = u \cos\theta,\ u_y = u \sin\theta,\ a_x = 0,\ a_y = -g$$

In Fig. (b),

$$u_x = u \cos\alpha,\ u_y = u \sin\alpha,\ a_x = -g \sin\beta,\ a_y = -g \cos\beta$$

Note *While solving the equations along x-axis, forget the motion along y-axis. Similarly, while taking the motion along y-axis, forget the motion along x-axis.*

Important Results Related to Projectile Motion

- List of formulae related to projectile motion

(i) $T = \dfrac{2u \sin\theta}{g} = \dfrac{2u_y}{g}$. (ii) $H = \dfrac{u^2 \sin^2\theta}{2g} = \dfrac{u_y^2}{2g}$.

(iii) $R = \dfrac{u^2 \sin 2\theta}{g} = u_x T = \dfrac{2u_x u_y}{g}$

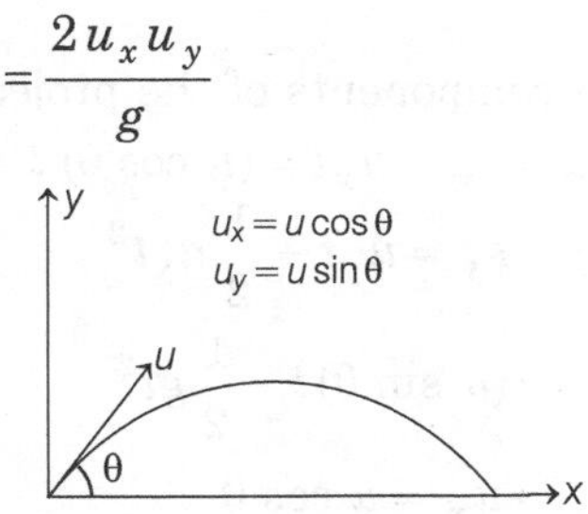

(iv) $R_{max} = \dfrac{u^2}{g}$ at $\theta = 45°$.

(v) $R_\theta = R_{90°-\theta}$ for same value of u.

(vi) Equation of trajectory,

$$y = x\tan\theta - \frac{gx^2}{2u^2\cos^2\theta}$$

$$= x\tan\theta - \frac{gx^2}{2u^2}(1+\tan^2\theta)$$

$$= x\left(1-\frac{x}{R}\right)\tan\theta$$

- Formulae of T, H and R can be applied directly between two points lying on same horizontal line.

  For example, in the two projectile motions shown in figure,

$$t_{OQM} = T = \frac{2u\sin\alpha}{g},\ PQ = H = \frac{u^2\sin^2\alpha}{2g}$$

and
$$OM = R = \frac{u^2\sin 2\alpha}{g}$$

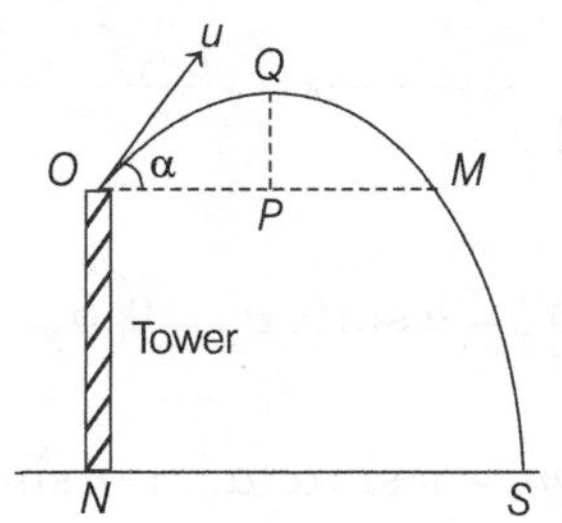

For finding t_{OQMS} or distance NS method-2 is applied.

- Since $a_x = 0$, motion of the projectile in horizontal direction is uniform.

  Hence, horizontal component of velocity $u \cos\theta$ does not change during its motion.
- Motion in vertical direction is first retarded, then accelerated in opposite direction because u_y is upwards and a_y is downwards.

  Hence, vertical component of its velocity first decreases, then increases in downward direction.
- The coordinates and velocity components of the projectile at time t are

$$x = s_x = u_x t = (u\cos\theta)\,t$$

$$y = s_y = u_y t + \frac{1}{2}a_y t^2$$

$$= (u\sin\theta)\,t - \frac{1}{2}gt^2$$

$$v_x = u_x = u\cos\theta$$

and
$$v_y = u_y + a_y t = u\sin\theta - gt$$

Therefore, speed of projectile at time t is $v = \sqrt{v_x^2 + v_y^2}$ and the angle made by its velocity vector with positive x-axis is

$$\theta = \tan^{-1}\left(\frac{v_y}{v_x}\right)$$

- In projectile motion, speed (and hence, kinetic energy) is minimum at highest point. Speed = (cos θ) times of the speed of projection and kinetic energy = ($\cos^2 \theta$) times of the initial kinetic energy. Here, θ = angle of projection.
- At a height difference of h, $v_x = u_x$, $v_y = \sqrt{u_y^2 \pm 2gh}$

  Take $+2gh$, if u_y is the value at some higher point and v_y is the value at some lower point and take $-2gh$ in opposite case.

When Projectile is Projected Horizontally

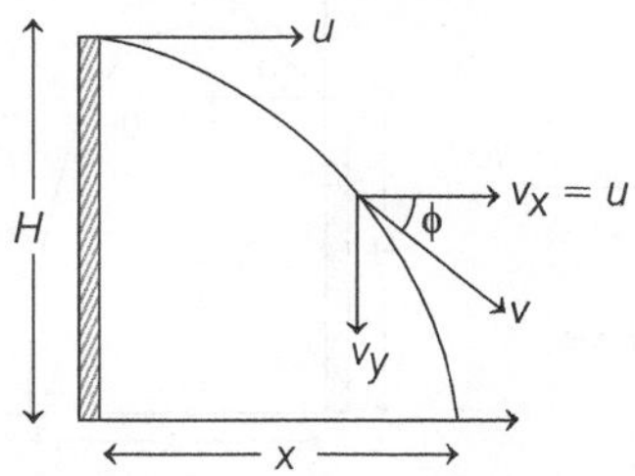

Initial velocity in vertical direction = 0

Time of flight, $T = \sqrt{\frac{2H}{g}}$

Horizontal range, $x = uT = u\sqrt{\frac{2H}{g}}$

Vertical velocity after t seconds, $v_y = gt$ $(\because u_y = 0)$

Velocity of projectile after t seconds, $v = \sqrt{v_x^2 + v_y^2} = \sqrt{u^2 + (gt)^2}$

If velocity makes an angle ϕ from horizontal, then $\tan \phi = \frac{v_y}{v_x} = \frac{gt}{u}$

When Projectile is Projected Downward at an Angle with Horizontal

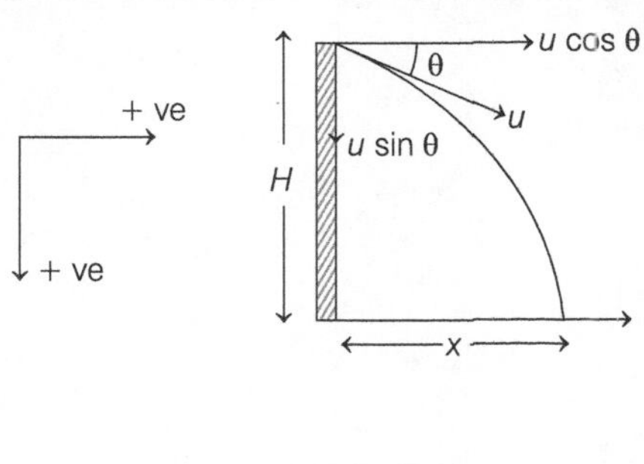

Initial velocity in horizontal direction = $u \cos \theta$
Initial velocity in vertical direction = $u \sin \theta$
Time of flight can be obtained from the equation,

$$H = (u \sin \theta) t + \frac{1}{2} g t^2$$

Horizontal range, $x = (u \cos \theta)\, t$
Vertical velocity after t second,

$$v_y = u \sin \theta + gt$$

Velocity of projectile after t second,

$$v = \sqrt{v_x^2 + v_y^2} = \sqrt{(u \cos \theta)^2 + (u \sin \theta + gt)^2}$$
$$= \sqrt{u^2 + (gt)^2 + 2ugt \sin \theta}$$

When Projectile is Projected Upward at an Angle with Horizontal

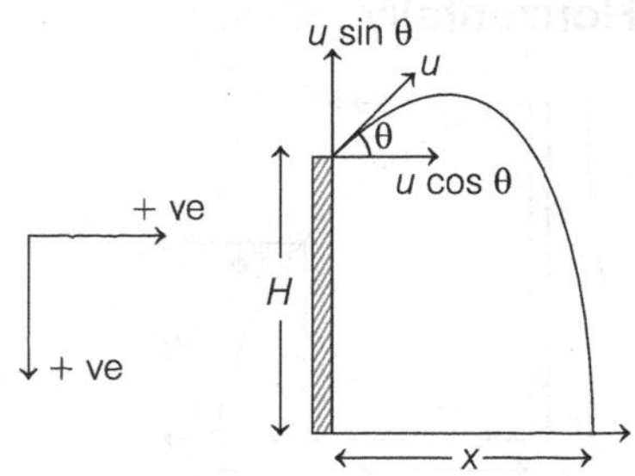

Initial velocity in horizontal direction = $u \cos \theta$
Initial velocity in vertical direction = $- u \sin \theta$
Time of flight can be obtained from the equation

$$H = (- u \sin \theta)\, t + \frac{1}{2} g t^2$$

Horizontal range, $x = (u \cos \theta) t$
Vertical velocity after t second, $v_y = (- u \sin \theta) + gt$
Velocity of projectile after t second

$$v = \sqrt{v_x^2 + v_y^2} = \sqrt{u^2 + (gt - u \sin \theta)^2}$$
$$= \sqrt{u^2 + (gt)^2 - 2ugt \sin \theta}$$

CHAPTER 04

Laws of Motion

Equilibrium of Forces

- **Concurrent coplanar forces** If all forces are in equilibrium or $\mathbf{F}_1 + \mathbf{F}_2 + \mathbf{F}_3 + \mathbf{F}_4 = 0$, then we can write

$$\Sigma F_x = 0 \text{ and } \Sigma F_y = 0$$

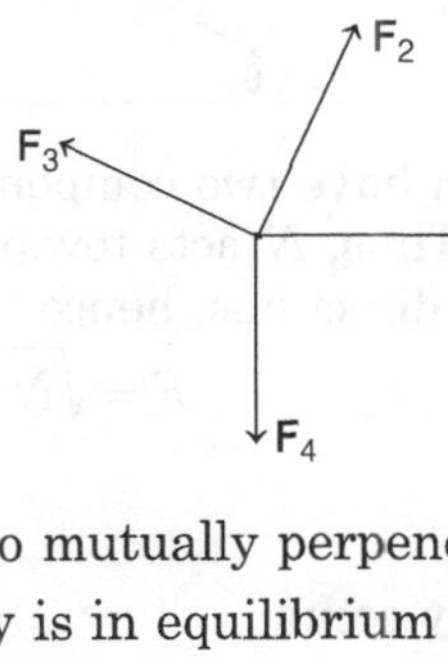

where, x and y are any two mutually perpendicular directions.

- **Lami's theorem** If a body is in equilibrium under three concurrent forces as shown in figure.

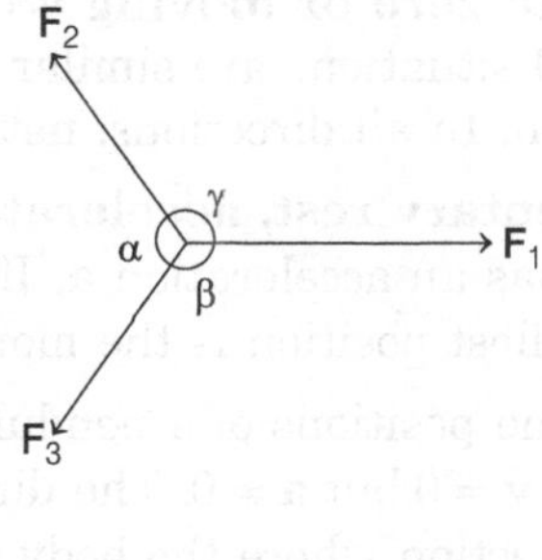

Then, we can write

$$\frac{F_1}{\sin\alpha} = \frac{F_2}{\sin\beta} = \frac{F_3}{\sin\gamma}$$

- **Non concurrent coplanar forces** If a body is in equilibrium under non concurrent coplanar forces, then we can write $\Sigma F_x = 0, \Sigma F_y = 0$ and Σ(moment about any point) = 0

Important Components

- Component of weight mg along the plane is $mg\sin\theta$ and component perpendicular to plane is $mg\cos\theta$.

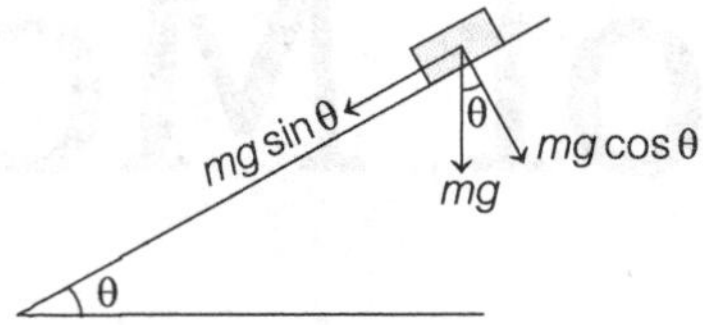

- If an inclined plane makes an angle θ with horizontal, then normal reaction N acting on a block kept over this surface makes θ with vertical. So, this normal reaction has vertical component $N\cos\theta$ and the horizontal component $N\sin\theta$.

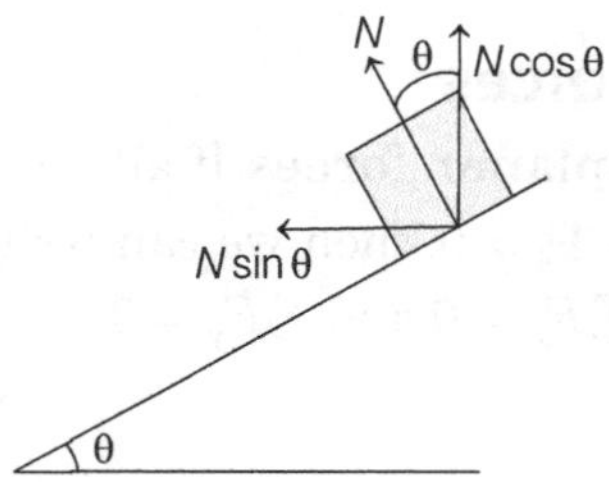

- Net contact force (F) can have two components, **friction** (f) which is tangential and **normal reaction** (N). Thus, N acts towards the body. Further N and f are in mutually perpendicular directions, hence

$$F = \sqrt{N^2 + f^2}$$

Resolution of Forces

Two situations normally occur :

(i) **Permanent rest, equilibrium, net force equal to zero, net acceleration equal to zero or moving with constant velocity** From forces point of view all situations are similar. In this case, we can resolve the forces in any direction. In all directions, net force should be zero.

(ii) **Momentary rest, accelerated or rotating in a circle** In all these cases, body has an acceleration **a**. If a particle is projected vertically upwards, then its highest position is the momentary rest position, where $\mathbf{a} \neq 0$ but $\mathbf{v} = 0$.

Extreme positions of a pendulum are also the momentary rest positions, where $\mathbf{v} = 0$ but $\mathbf{a} \neq 0$. The direction of **a** in the momentary rest condition is the direction where the body will move just after few seconds. In a uniform circular motion, direction of **a** is towards centre.

In the above situations, we normally resolve the forces in the direction of acceleration and perpendicular to it. In the direction of acceleration, net force should be $m\mathbf{a}$ and perpendicular to acceleration, net force should be zero.

Note *In the above situations, we can also resolve the forces in any third direction (say x-direction). In that case, net force along x-direction = m (component of* ***a*** *along x-direction)*

- If **a** is the acceleration of a body, then $m\mathbf{a}$ force does not act on the body but this much force is required to provide **a** acceleration to the body. The different available forces acting on the body provide this $m\mathbf{a}$ force or we can say that vector sum of all forces acting on the body is equal to $m\mathbf{a}$.

  The available forces may be weight, tension, normal reaction, friction or any externally applied force, etc.

- If all bodies of a system have a common acceleration, then that common acceleration can be given by

$$a = \frac{\text{net pulling/ pusing force}}{\text{total mass}} = \frac{\text{NPF}}{\text{TM}}$$

  Net pulling/pushing force (NPF) is actually the net force.

  After finding that common acceleration, we will have to draw free body diagrams of different blocks to find normal reaction or tension, etc.

- In some cases, acceleration of a block is inversely proportional to tension force acting on the block (or its component in the direction of motion or acceleration). If tension is double (as compared to other block), then acceleration will be half.

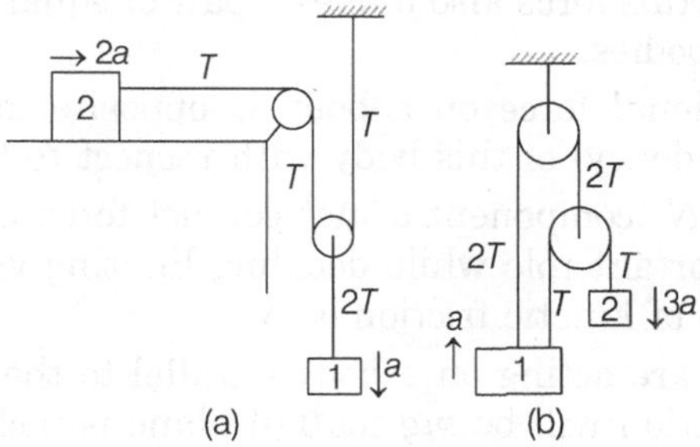

In Fig. (a), tension force on block-1 is double ($=2T$) the tension force on block-2 ($=T$). Therefore, acceleration of block-1 will be half.

If block-1 has an acceleration a in downward direction, then block-2 will have an acceleration $2a$ towards right.

In Fig. (b), tension force on block-1 is three times ($2T + T = 3T$), the tension force on block-2 ($=T$). Therefore, acceleration of block-2 will be three times. If block-1 has an acceleration a in upward direction, then acceleration of block-2 will be $3a$ downwards.

Pseudo Force

If we observe an object from a non-inertial frame, then the motion conditions are changed from this frame. To justify these changed motion conditions from equation point of view, we will have to apply a pseudo force ma in the opposite direction of acceleration of frame.

In the pseudo force, mass m is the mass of object which is being observed and a is the acceleration of frame from where this object is observed.

Inertial Frame of Reference

A non-accelerating frame of reference is called an inertial frame of reference. A frame of reference moving with a constant velocity is an inertial frame of reference.

Non-inertial Frame of Reference

An accelerating frame of reference is called a non-inertial frame of reference.

Note *(i) A rotating frame of reference is a non-inertial frame of reference, because it is also an accelerating reference.*

(ii) Earth is rotating about its axis of rotation and it is revolving around the centre of sun also. So, it is a non-inertial frame of reference. But for most of the cases, we consider it as an inertial frame of reference.

Friction

- Friction is a tangential component of net contact force between two solid bodies in contact.
- This force starts acting between them when there is relative motion (or have the tendency of relative motion).
- Like other forces, this force also makes a pair of equal and opposite forces acting on two different bodies.
- Direction of frictional force on a body is opposite to the direction of relative motion (or its tendency) of this body with respect to the other body.
- Normal reaction N (component of net contact force in perpendicular direction) plays a very important role while deciding limiting value of static friction $\mu_s N$ or constant value of kinetic friction $\mu_k N$.
- As long as forces are acting on a body parallel to the plane over which body is kept, normal reaction will be $mg \cos\theta$ (if plane is inclined).

  If forces are acting at some angle with plane, normal reaction is greater than mg (or $mg \cos\theta$) or less than this, depending upon whether the external forces are of pushing nature or pulling nature.

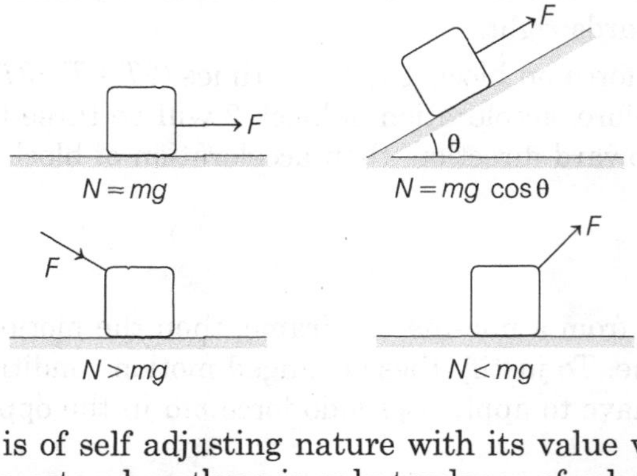

- Static friction is of self adjusting nature with its value varying between 0 and $\mu_s N$. This force acts when there is only tendency of relative motion.

  On the other hand, kinetic friction is constant of value $\mu_k N$. This force acts when relative motion actually takes place.

- Coefficient of kinetic friction (μ_k) is always less than the coefficient of static friction (μ_s).
- If μ_s and μ_k are not separately given, but only one value of μ is given, then in this case, limiting value of static friction and constant value of kinetic friction are same and equal to μN.
- 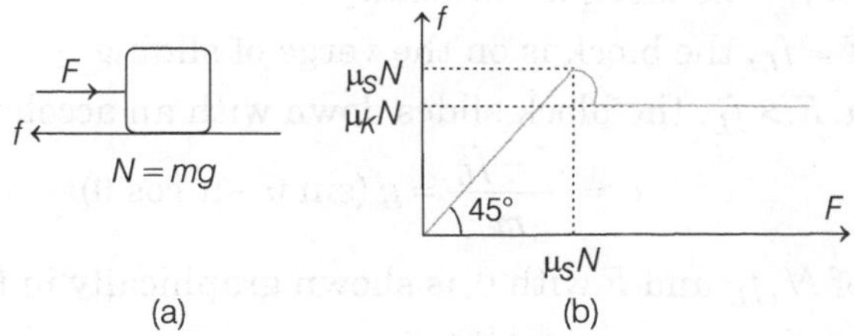

(a) (b)

In Fig. (a), a force F is applied on a block of mass m. Force of friction f starts acting on the block in opposite direction to stop its relative motion with ground. Following cases may arise depending on the value of F.

If $F \le \mu_s N, f = F, F_{\text{net}} = 0 \therefore \quad a = 0$

If $F > \mu_s N, f = \mu_k N, F_{\text{net}} = F - f$ and $a = \dfrac{F_{\text{net}}}{m}$

Corresponding graph is shown in Fig. (b).

Angle of Friction (λ)

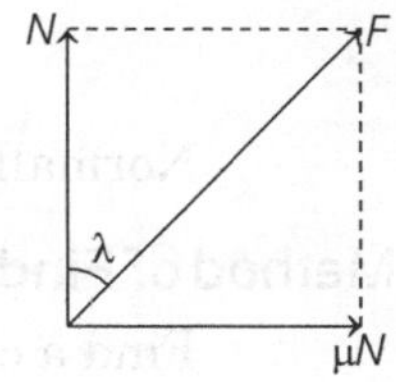

In critical condition when slipping is about to occur, the two forces acting are the normal reaction N and frictional force μN.

The resultant of these two forces is F and it makes an angle λ with the normal reaction, where

$$\tan \lambda = \frac{\mu N}{N} = \mu$$

or

$$\lambda = \tan^{-1}(\mu)$$

This angle λ is called the angle of friction.

Angle of Repose (α)

Suppose a block of mass m is placed on an inclined plane whose inclination θ can be increased or decreased. Let μ be the coefficient of friction between the block and the plane.

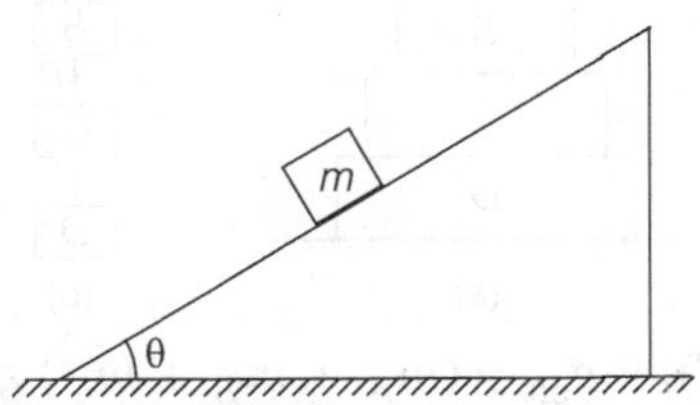

At a general angle θ,

$N = mg\cos\theta$, $f_L = \mu N = \mu mg\cos\theta$ and downward force $F = mg\sin\theta$.

As θ increases, F increases and f_L decreases. At angle $\theta = \alpha$, called angle of repose, $F = f_L$ and the block starts sliding.

$$\therefore \quad mg \sin \alpha = \mu mg \cos \alpha$$

$$\text{or} \quad \tan \alpha = \mu \quad \text{or } \alpha = \tan^{-1} (\mu)$$

If $\theta < \alpha$, $F < f_L$, the block is stationary.

If $\theta = \alpha$, $F = f_L$, the block is on the verge of sliding.

and if $\theta > \alpha, F > f_L$, the block slides down with an acceleration

$$a = \frac{F - f_L}{m} = g (\sin \theta - \mu \cos \theta)$$

Variation of N, f_L and F with θ, is shown graphically in figure.

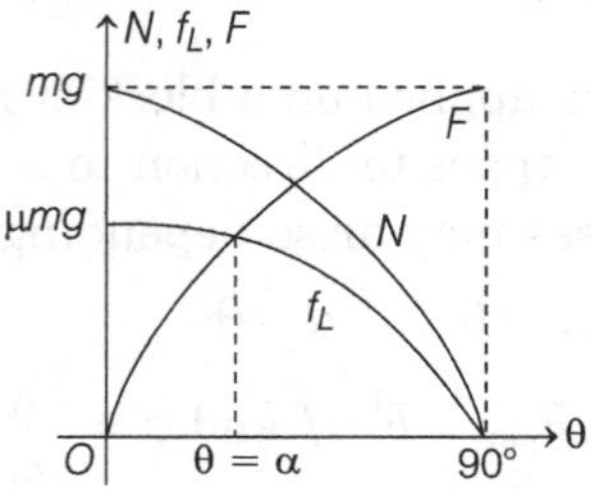

$$N = mg \cos \theta \quad \text{or} \quad N \propto \cos \theta$$

$$f_L = \mu mg \cos \theta \quad \text{or} \quad f_L \propto \cos \theta$$

$$F = mg \sin \theta \quad \text{or} \quad F \propto \sin \theta$$

Normally $\mu < 1$, so $f_L < N$.

Method of Finding Tension at Some Point (say P), if it is Non-uniform

Find a common acceleration a of all the blocks attached with the given string. In some cases, a will be given in the question. Cut the string at P and make free body diagram of any one part.

In addition to other forces which are actually acting on this part, apply one tension at P. Direction of this tension should be such that string looks as if it is stretched. Now apply, $F_{net} = ma$ for this part to find the tension at P.

- In Fig. (a), normal reaction at point P (between blocks C and D) is given by

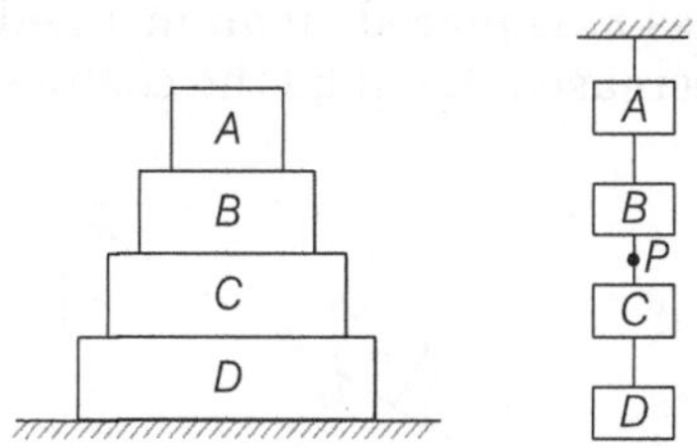

(a) (b)

$N = [\Sigma(\text{mass above } P)] \times g_{\text{eff}} = (m_A + m_B + m_C)\, g_{\text{eff}}$

In Fig. (b), tension at point P is given by

$$T = \Sigma\ [(\text{mass above } P)] \times g_{\text{eff}} = (m_A + m_B) g_{\text{eff}}$$

If the strings are massless, then $T = (m_C + m_D) g_{\text{eff}}$

Here, $g_{eff} = g$ if acceleration of system is zero
$= (g + a)$ if acceleration a is upwards
$= (g - a)$ if acceleration a is downwards

Few Importants Points Related to Newton's Laws of Motion

- **Feeling of weight to a person is due to the normal reaction.** Under normal conditions, $N = mg$, feeling of weight is the actual weight mg. If we are standing on a lift and the lift has an acceleration a upwards then $N = m(g + a)$. Therefore, feeling of weight is more than the actual weight mg.
  Similarly, if a is downwards, then $N = m(g - a)$ and feeling of weight is less than the actual weight mg.
- If a car (or any other vehicle) accelerates and decelerates by friction, then the maximum acceleration or deceleration of a car on horizontal ground can be $\mu g\left(=\frac{f_L}{m}=\frac{\mu N}{m}=\frac{\mu mg}{m}\right)$, unless some other force is applied.
- Kilogram weight (kg-wt) or kilogram force (kg-f), gram weight (g-wt) or gram-force (g-f) are also the units of force. The CGS unit of force is dyne.
$$1\text{ N} = 10^5\text{ dyne}$$
$$1\text{ kg-wt} = 1\text{ kg-f} = 9.81\text{ N}$$
$$1\text{ g-wt} = 1\text{ g-f} = 981\text{ dyne}$$
- Newton's second law of motion is called real law of motion because first and third laws of motion can be obtained by it.
- Newton's third law of motion is a consequence of law of conservation of momentum. Rocket propulsion is based on law of conservation of momentum or Newton's third law of motion.
- Kinetic friction is less than the limiting value of static friction. This is because, to start motion of a body molecular bonds formed between two bodies in contact are broken. But once the motion starts, new bonds are formed and broken and this process continues.

Examples of Newton's First Law

(i) When a carpet or a blanket is beaten with a stick, then the dust particles separate out from it.

(ii) If a moving vehicle suddenly stops, then the passengers inside the vehicle bend in forward direction.

Examples of Newton's Third Law

(i) Swimming becomes possible because of third law of motion.

(ii) Jumping of a man from a boat onto the bank of a river.

(iii) Jerk is produced in a gun when bullet is fired from it.

(iv) Pulling of cart by a horse.

CHAPTER 05

Work, Power and Energy

Work Done

- **By a constant force**

  $W = \mathbf{F} \cdot \mathbf{s} = \mathbf{F} \cdot (\mathbf{r}_f - \mathbf{r}_i) = Fs\cos\theta$

  = Force × displacement in the direction of force.

  = Displacement × component of force in the direction of displacement.

  = + **Fs** if $\theta = 0°$

  = − **Fs** if $\theta = 180°$

  = 0 if $\theta = 90°$, where θ is the angle between **F** and **s**.

- **By a variable force** $W = \int_{x_i}^{x_f} F \cdot x$, where $F = f(x)$
- **By area under F - x graph** If force is a function of x, we can find work done by area under F - x graph with projection along x-axis. In this method, magnitude of work done can be obtained by area under F-x graph, but sign of work done should be decided by you.

  If force and displacement both are positive or negative, work done will be positive. If one is positive and other is negative, then work done will be negative.

- Work done by a force may be positive, negative or zero, depending upon the angle (θ) between the force vector **F** and displacement vector **s**. Work done by a force is zero when $\theta = 90°$, it is positive when $0° \le \theta < 90°$ and negative when $90° < \theta \le 180°$.

  For example, when a person lifts a body, the work done by the lifting force is positive (as $\theta = 0°$) but work done by the force of gravity is negative (as $\theta = 180°$).

- Work depends on frame of reference. For example, if a person is pushing a box inside a moving train, then work done as seen from the frame of reference of train is $\mathbf{F} \cdot \mathbf{s}$ while as seen from the ground it is $\mathbf{F} \cdot (\mathbf{s} + \mathbf{s}_0)$. Here $\mathbf{s}_0$, is the displacement of train relative to ground.

Note $1 \text{ J} = 10^7$ erg

Conservative and Non-Conservative Forces

- In case of conservative forces, work done is path independent.
- Potential energy is defined only for conservative forces.
- If only conservative forces are acting on a system, its mechanical energy should remain constant.
- In case of non-conservative forces, work done is dependent upon the path.
- Examples of conservative forces are gravitational force between two point masses, electrostatic force between two charges and the spring force.
- Examples of non-conservative forces are force of friction, viscous force etc.

Potential Energy

- Potential energy is defined only for conservative forces.
- In a conservative force field, difference in potential energy between two points is the negative of work done by conservative forces in displacing the body (or system) from some initial position to final position. Hence,

$$\Delta U = -W$$

or $$U_B - U_A = -W_{A \to B}$$

- Change in potential energy is equal to the negative of work done by the conservative force ($\Delta U = -\Delta W$). If work done by the conservative force is negative, change in potential energy will be positive or potential energy of the system will increase and *vice-versa.*

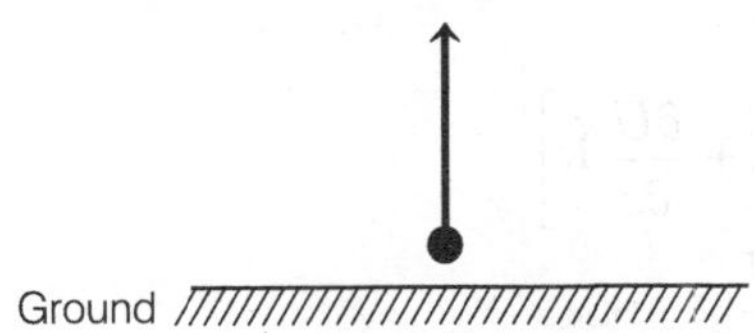

This can be explained by a simple example. Suppose a ball is taken from the ground to some height, work done by gravity is negative, i.e. change in potential energy should increase or potential energy of the ball will increase.

$$\Delta W_{gravity} = -\text{ve}$$

$\therefore$ $$\Delta U = +\text{ve} \qquad (\because \Delta U = -\Delta W)$$

or $$U_f - U_i = +\text{ve}$$

- Absolute potential energy at a point can be defined with respect to a reference point, where potential energy is assumed to be zero.

Reference point corresponding to gravitational potential energy and electrostatic potential energy is assumed at infinity.

Reference point corresponding to spring potential energy is taken at a point at natural length of spring.

Now, negative of work done in displacing the body from reference point (say O) to the point under consideration (say P) is called absolute potential energy at P. Thus,

$$U_P = -W_{O \to P}$$

- For increasing or decreasing in gravitational potential energy of a particle (for small heights) we can write,

$$\Delta U = \pm mgh$$

Here, h is the difference in heights of particle. In case of a rigid body, h of centre of mass of the rigid body is seen.

- $F = -\frac{dU}{dr}$, i.e. conservative forces always act in a direction, where potential energy of the system is decreased. This can also be shown as in figure.

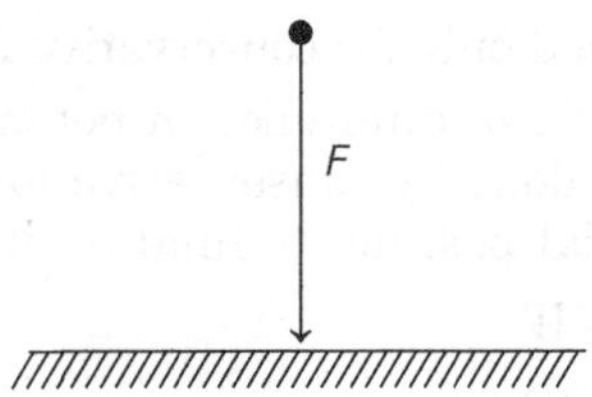

If a ball is dropped from a certain height. The force on it (its weight) acts in a direction in which its potential energy decreases.

Relation between Conservative Force (F) and its Potential Energy (*U*)

Conversion of *U* into F

- $\mathbf{F} = -\left[\frac{\partial U}{\partial x}\hat{\mathbf{i}} + \frac{\partial U}{\partial y}\hat{\mathbf{j}} + \frac{\partial U}{\partial z}\hat{\mathbf{k}}\right]$
- $F = -\frac{dU}{dr}$ or $-\frac{dU}{dx}$

  or $-$ (slope of U - r or U - x graph)

Conversion of F into *U*

- $dU = -\mathbf{F} \cdot d\mathbf{r}$

  Here, $\mathbf{F}$ will be given in the question and $d\mathbf{r} = (dx\hat{\mathbf{i}} + dy\hat{\mathbf{j}} + dz\hat{\mathbf{k}})$ is a standard vector.
- $dU = -Fdr$ or $-Fdx$

Spring Force

- $F = -kx$ or $F \propto -x$

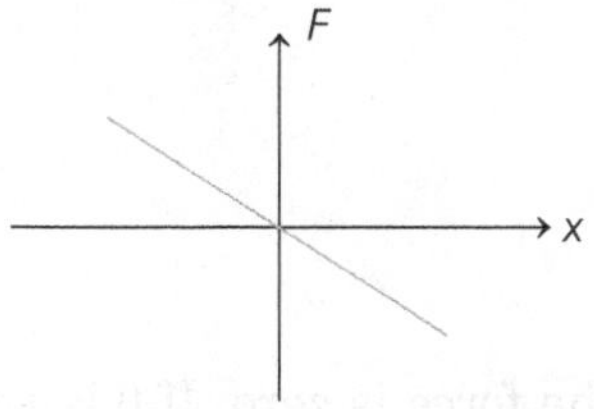

- $U_x = \frac{1}{2}kx^2$

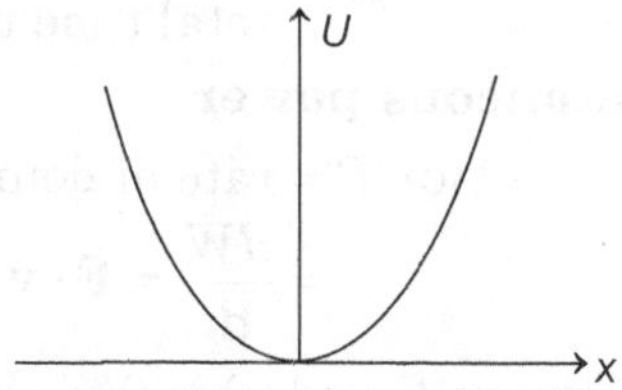

Work Done and Energy

- Work done by conservative forces is equal to minus of change in potential energy,
$$W_c = -\Delta U = -(U_f - U_i) = U_i - U_f$$
- Work done by all the forces is equal to change in kinetic energy,
$$W_{\text{all}} = \Delta K = K_f - K_i \qquad \text{(Work-energy theorem)}$$
- Work done by the forces other than the conservative forces (non-conservative + external forces) is equal to change in mechanical energy
$$W_{nc} + W_{\text{ext}} = \Delta E = E_f - E_i = (K_f + U_f) - (K_i + U_i)$$
- If there are no non-conservative forces, then
$$W_{\text{ext}} = \Delta E = E_f - E_i$$
Further, in this case, if no information is given regarding the change in kinetic energy, then we can take it zero. In that case,
$$W_{\text{ext}} = \Delta U = U_f - U_i$$

Types of Equilibrium

Physical situation	Stable equilibrium	Unstable equilibrium	Neutral equilibrium
Net force	Zero	Zero	Zero
Potential energy	Minimum	Maximum	Constant
When displaced from mean (equilibrium) position.	A restoring nature of force will act on the body, which brings the body back towards mean position.	A force will act which moves the body away from mean position.	Force is again zero.
In F - r graph 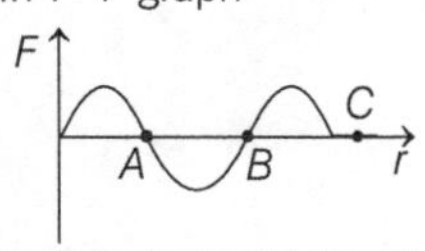	At point A	At point B	At point C
In U - r graph 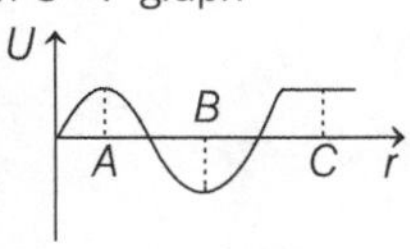	At point B	At point A	At point C

Power

- **Average power**

$$P_{av} = \frac{\text{total work done}}{\text{total time taken}} = \frac{W_{total}}{t}$$

- **Instantaneous power**

$$P_i \text{ or } P = \text{rate of doing work done}$$
$$= \frac{dW}{dt} = \mathbf{F} \cdot \mathbf{v} = Fv\cos\theta$$

- If θ (between **F** and **v**) is 90°, then power of the force is zero. If θ is acute, then power is positive and if θ is obtuse, then power is negative.

Some Important Points Related to Work and Energy

- Suppose a particle is released from point A with $u = 0$.

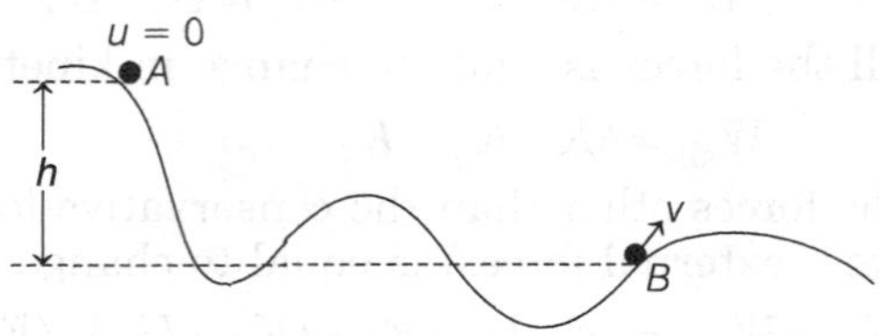

If friction is absent everywhere, then velocity at B will be

$v = \sqrt{2gh}$ (Irrespective of the track it follows from A to B)

Here, $h = h_A - h_B$

- In circular motion, centripetal force acts towards the centre. This force is perpendicular to small displacement $d\mathbf{s}$ and velocity **v**. Hence, work done by it is zero and power of this force is also zero.
- If friction is absent, then

$E_i = E_f$ (where, E = mechanical energy)

If friction exists, then

$$E_i - E_f = \text{WDAF}$$

Here, WDAF = work done against friction

$= \mu N d$

where, μ = coefficient of friction

N = normal reaction

and d = distance travelled on rough surface.

Note *In most of the cases,*
$N = mg$ on horizontal ground and
$N = mg\cos\theta$ on inclined ground

- Work done by friction $= E_f - E_i = -(\text{WDAF})$

CHAPTER 06

Circular Motion

Terms Related to Circular Motion

- A particle in circular motion may have two types of velocities :
  (i) linear velocity v and (ii) angular velocity ω

  These two velocities are related by the equation

  $$v = R\omega \quad \text{(where, } R = \text{radius of circular path)}$$

- Acceleration of particle in circular motion may have two components:
  (i) tangential component (a_t) and
  (ii) normal or radial component (a_n)

  As the name suggests, tangential component is tangential to the circular path, given by

  $$a_t = \text{rate of change of speed} = \frac{dv}{dt} = \frac{d\,|\mathbf{v}|}{dt} = R\alpha$$

  where, α = angular acceleration = rate of change of angular velocity

  $$= \frac{d\omega}{dt}$$

  The normal or radial component also known as centripetal acceleration is towards centre and is given by

  $$a_n = R\omega^2 = \frac{v^2}{R}$$

- Net acceleration of particle is resultant of two perpendicular components a_n and a_t. Hence,

  $$a = \sqrt{a_n^2 + a_t^2}$$

- Tangential component a_t is responsible for change of speed of the particle. This can be positive, negative or zero, depending upon the situation whether the speed of particle is increasing, decreasing or constant.

  Normal component is responsible for change in direction of velocity. This component can never be equal to zero in circular motion.

- In general, in any curvilinear motion, direction of instantaneous velocity is tangential to the path, while acceleration may have any direction.

If we resolve the acceleration in two normal directions, one parallel to velocity and another perpendicular to velocity, the first component is a_t while the other is a_n.

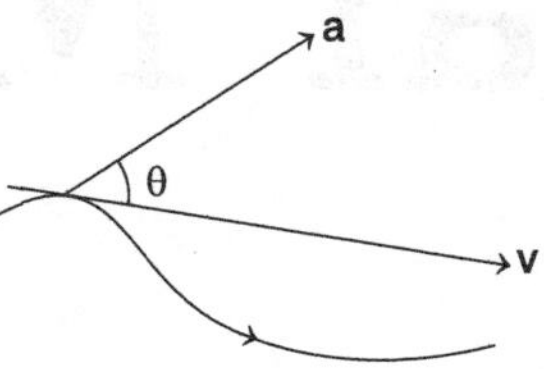

Thus, a_t = component of **a** parallel to **v**

$$= a\cos\theta = \frac{\mathbf{a}\cdot\mathbf{v}}{v}$$

$$= \frac{dv}{dt} = \frac{d|\mathbf{v}|}{dt} = \text{rate of change of speed.}$$

and a_n = component of **a** perpendicular to **v**

$$= \sqrt{a^2 - a_t^2} = v^2/R$$

Here, v is the speed of particle at that instant and R is called the radius of curvature to the curvilinear path at that point.

- In $a_t = a\cos\theta$, if θ is acute, a_t will be positive and speed will increase. If θ is obtuse, a_t will be negative and speed will decrease. If θ is 90°, a_t is zero and speed will remain constant.
- Now, depending upon the value of a_t, circular motion may be of three types :
  (i) Uniform circular motion in which speed remains constant or $a_t = 0$.
  (ii) Circular motion of increasing speed in which a_t is positive.
  (iii) Circular motion of decreasing speed in which a_t is negative.

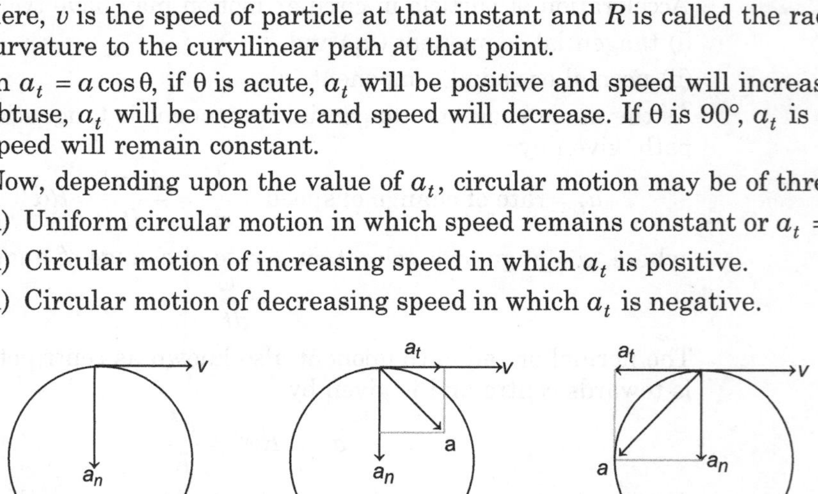

$a_t = 0$, $a = a_n$, v = constant

$a = \sqrt{a_n^2 + a_t^2}$, v is increasing

$a = \sqrt{a_n^2 + a_t^2}$, v is decreasing

- Circular motion is a two dimensional motion (motion in a plane).

Linear velocity vector and linear acceleration vector lie in the plane of circle. Angular velocity vector and angular acceleration vector are perpendicular to the plane of the circle given by right hand screw law.

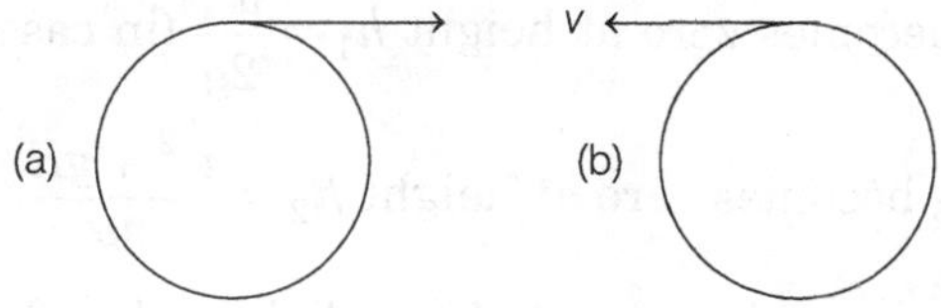

In Fig. (a), for example, if speed is increasing, then angular velocity vector and angular acceleration vector both are perpendicular to paper inwards.

In Fig. (b), if speed is decreasing, then angular velocity vector is perpendicular to paper inwards while angular acceleration vector is perpendicular to paper outwards.

Vertical Circular Motion

- Suppose a bob of mass m is suspended from a light string of length R as shown. If velocity at bottommost point of bob is u, then depending upon the value of u following three cases are possible :

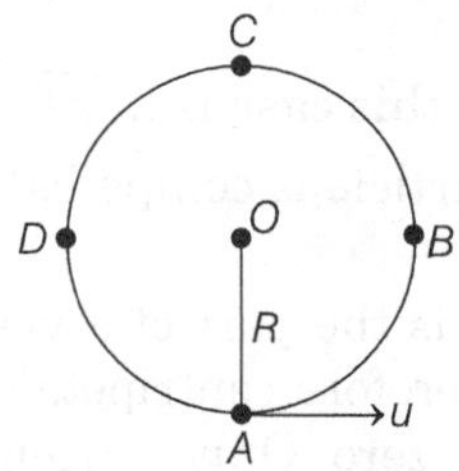

(i) If $u \geq \sqrt{5gR}$, bob will complete the circle.

(ii) If $\sqrt{2gR} < u < \sqrt{5gR}$, string will slack between B and C. At the time of slacking, tension in the string will become zero ($T = 0$) and velocity is non-zero ($v \neq 0$). After slacking motion of bob is projectile.

(iii) If $u \leq \sqrt{2gR}$, bob will oscillate between BAD. In this case, $v = 0$ but $T \neq 0$.

The above three conditions have been derived for a particle moving in a vertical circle attached to a string.

- The same conditions apply, if a particle moves inside a smooth spherical shell of radius R. The only difference is that the tension is replaced by the normal reaction N.
- If $u = \sqrt{5gR}$, bob will just complete the circle. In this case, velocity at topmost point is $v = \sqrt{gR}$. Tension in this critical case is zero at topmost point and 6 mg at bottommost point.
- If $u = \sqrt{2gR}$, bob will just reach to the point B. At that point, velocity and tension both will become zero.
- At height h from bottom, velocity of bob will be $v = \sqrt{u^2 - 2gh}$.

- Velocity of bob becomes zero at height $h_1 = \frac{u^2}{2g}$ (in case of oscillation) and tension in string becomes zero at height $h_2 = \frac{u^2 + gR}{3g}$.
- If a particle of mass m is connected to a light rod and whirled in a vertical circle of radius R, then to complete the circle, the minimum velocity of the particle at the bottommost point is not $\sqrt{5\,gR}$. This is because in this case, velocity of the particle at the topmost point can be zero also.

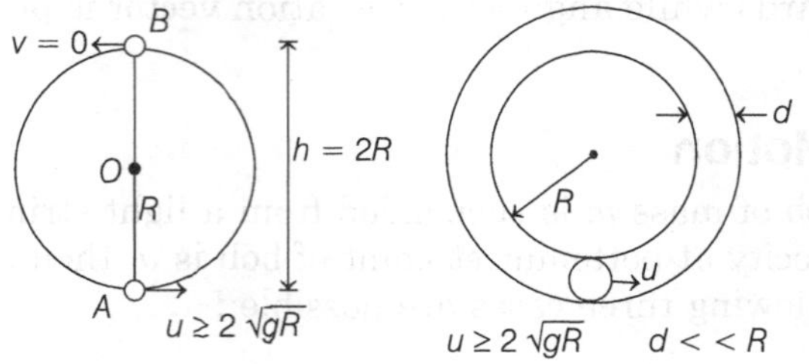

The minimum value of u in this case is $2\sqrt{gR}$ or $\sqrt{4gR}$.
Same is the case when a particle is compelled to move inside a smooth vertical tube.

- Oscillation of a pendulum is the part of a vertical circular motion. At points A and C, velocity is zero, therefore centripetal force or acceleration (also called radial acceleration) will be zero. Only tangential force or acceleration is present. From A to B or C to B, speed of the bob increases.

  Therefore, tangential force or acceleration is parallel to velocity. From B to A or B to C, speed of the bob decreases. Hence, tangential force or acceleration is antiparallel to velocity.

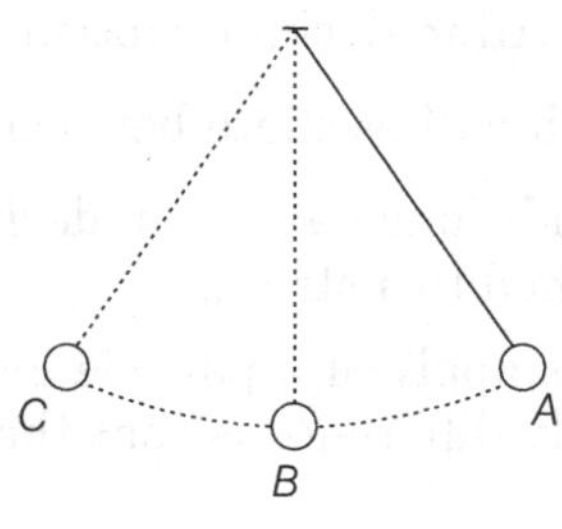

Dynamics of Circular Motion

- In circular motion, forces are normally resolved in tangential and radial directions. In tangential direction, net force should be ma_t and in radial direction, net force should be ma_r or $\frac{mv^2}{R}$ or $mR\omega^2$. In uniform circular motion, $a_t = 0$. Hence, in tangential direction, net force should be zero.
- If plane of uniform circular motion is horizontal, then one of the tangent is vertical also. So, we resolve the forces in horizontal and vertical directions.

In vertical tangential direction, net force is zero and in horizontal radial direction towards centre, net force should be $\frac{mv^2}{R}$ or $mR\omega^2$.

- In vertical circular motion, speed of particle continuously keeps on changing. As the particle moves upwards, speed or kinetic energy decreases. Since, speed is continuously changing, so a_t is never zero.
- During the motion only two forces tension (T) and weight (mg) are acting. Tension is already in radial direction towards centre. So, we will have to resolve only mg.

Condition of Toppling of a Vehicle on Circular Track

- While moving in a circular track, normal reaction on the outer wheels (N_1) is more than the normal reaction on inner wheels (N_2).

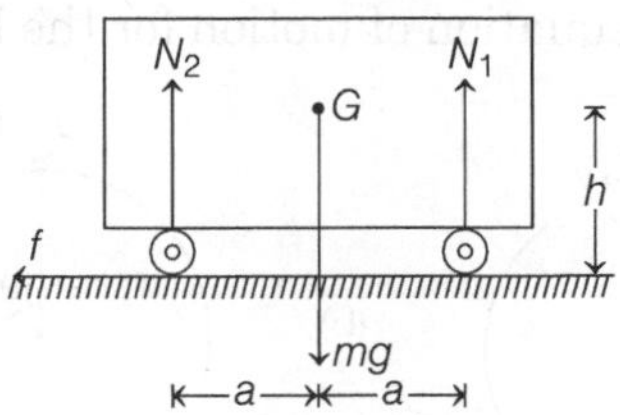

or $$N_1 > N_2$$

This can be proved as below.

$$N_1 + N_2 = mg \quad \text{...(i)}$$

and $$f = \frac{mv^2}{r} \quad \text{...(ii)}$$

For rotational equilibrium of car, net torque about centre of gravity should be zero.

or $$N_1(a) = N_2(a) + f(h) \quad \text{...(iii)}$$

$\Rightarrow$ $$N_2 = N_1 - \left(\frac{h}{a}\right) f = N_1 - \left(\frac{mv^2}{r}\right)\left(\frac{h}{a}\right) \quad \text{...(iv)}$$

or $$N_2 < N_1$$

From Eq. (iv), we see that N_2 decreases as v increases.

In critical case, $N_2 = 0$

and $$N_1 = mg \quad \text{[From Eq. (i)]}$$

$\therefore$ $$N_1(a) = f(h) \quad \text{[From Eq. (iii)]}$$

or $$(mg)(a) = \left(\frac{mv^2}{r}\right)(h)$$

or $$v = \frac{gra}{h}$$

Now, if $v > \sqrt{\frac{gra}{h}}$, $N_2 < 0$ and the car topples outwards.

Therefore, for a safe turn without toppling, $v \le \sqrt{\frac{gra}{h}}$.

- From the above discussion, we can conclude that while taking a turn on a level road there are two critical speeds, one is the maximum speed for sliding $(= \sqrt{\mu r g})$ and another is maximum speed for toppling $\left(= \sqrt{\frac{gra}{h}}\right)$. One should keep ones car's speed less than both for neither to slide nor to overturn.

Motion of a Ball over a Smooth Solid Sphere

Suppose a small ball of mass m is given a velocity v over the top of a smooth sphere of radius R. The equation of motion for the ball at the topmost point will be

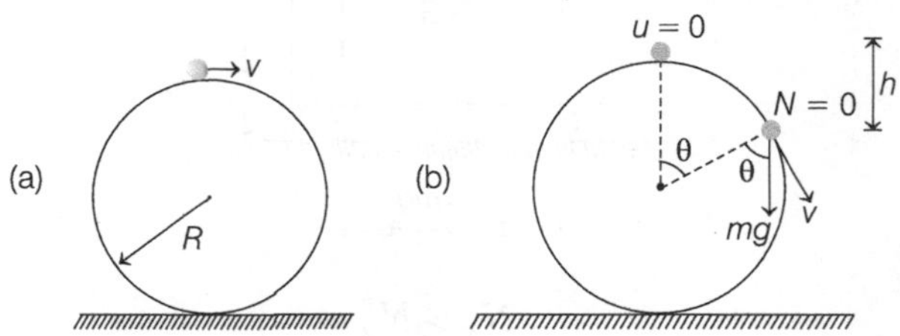

$$mg - N = \frac{mv^2}{R}$$

or
$$N = mg - \frac{mv^2}{R}$$

From this equation, we see that the value of N decreases as v increases. Minimum value of N can be zero.

Hence,
$$0 = mg - \frac{mv_{max}^2}{R}$$

or
$$v_{max} = \sqrt{Rg}$$

So, ball will lose contact with the sphere right from the beginning if velocity of the ball at topmost point $v > \sqrt{Rg}$.

If $v < \sqrt{Rg}$, it will lose contact after moving certain distance over the sphere.

Now, let us find the angle θ, where the ball loses contact with the sphere if velocity at topmost point is just zero. [see Fig. (b)].

$$h = R(1 - \cos\theta) \quad \text{...(i)}$$

$$v^2 = 2gh \quad \text{...(ii)}$$

$$mg\cos\theta = \frac{mv^2}{R} \quad (\text{as } N = 0) \quad \text{...(iii)}$$

Solving Eqs. (i), (ii) and (iii), we get

$$\theta = \cos^{-1}\left(\frac{2}{3}\right) = 48.2°$$

Thus, the ball can move on the sphere maximum upto $\theta = \cos^{-1}\left(\frac{2}{3}\right)$.

Other Cases of Circular Motion

- In the following two figures, surface is smooth. So, only two forces N and mg are acting. But directions of acceleration are different.

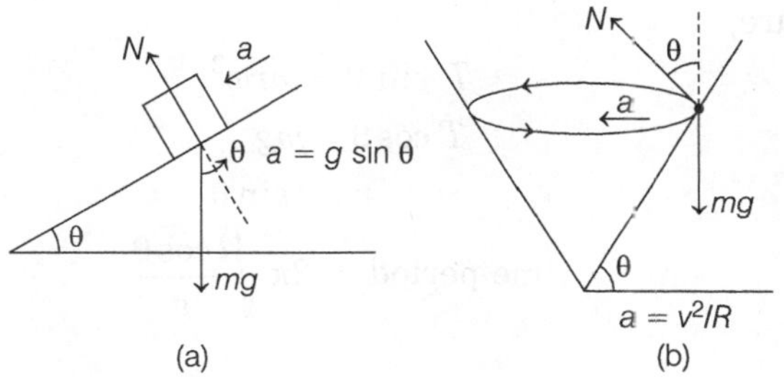

Net force perpendicular to acceleration should be zero. So, in the first figure,

$$N = mg\cos\theta$$

and in the second figure, $N\cos\theta = mg$

- (i) When a vehicle is moving over a convex bridge, then at the maximum height, reaction (N_1) is

$$N_1 = mg - \frac{mv^2}{r}$$

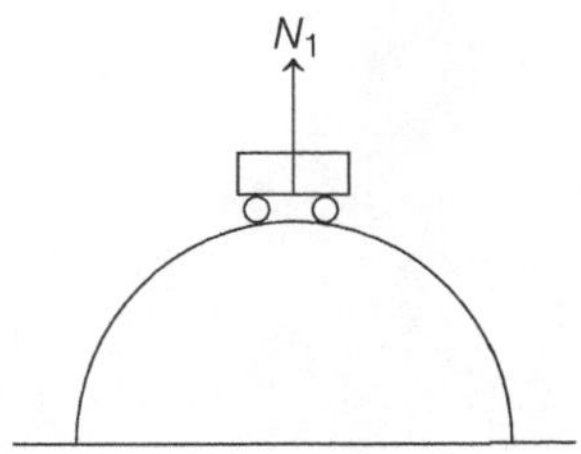

(ii) When a vehicle is moving over a concave bridge, then at the lowest point, reaction (N_2) is

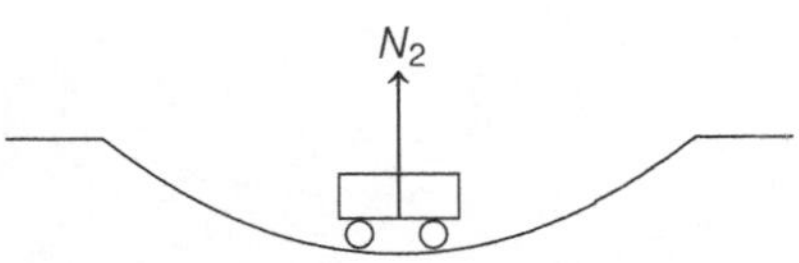

$$N_2 = mg + \frac{mv^2}{r}$$

- **Conical Pendulum** If a simple pendulum is fixed at one end and the bob is rotating in a horizontal circle, then it is called a conical pendulum.

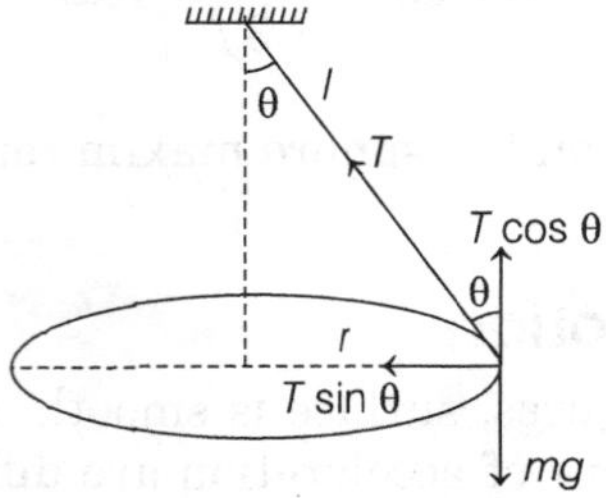

From the figure,

$$T \sin \theta = mr\omega^2$$

$$T \cos \theta = mg$$

$$r = l \sin \theta$$

and time period $= 2\pi \sqrt{\dfrac{l \cos \theta}{g}}$

CHAPTER 07

Centre of Mass, Momentum and Impulse

Position of Centre of Mass

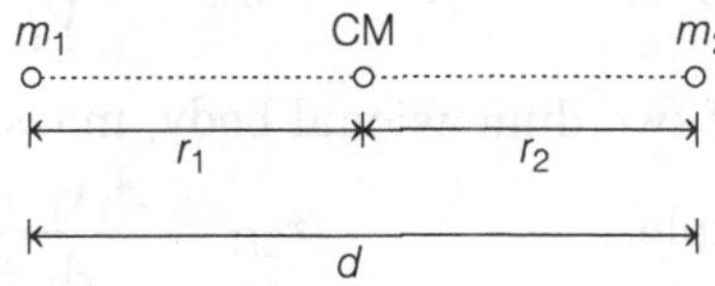

- **Two point masses**

$$r \propto \frac{1}{m} \quad \text{or} \quad \frac{r_1}{r_2} = \frac{m_2}{m_1}$$

$$\therefore \quad m_1 r_1 = m_2 r_2$$

$$\Rightarrow \quad r_1 = \frac{m_2}{m_1 + m_2} \cdot d$$

$$\Rightarrow \quad r_2 = \frac{m_1}{m_1 + m_2} \cdot d$$

- **More than two point masses**

(i) $\mathbf{r}_{CM} = \dfrac{m_1 \mathbf{r}_1 + m_2 \mathbf{r}_2}{m_1 + m_2}$

(ii) $X_{CM} = \dfrac{m_1 x_1 + m_2 x_2}{m_1 + m_2}$

$$\Rightarrow \quad Y_{CM} = \frac{m_1 y_1 + m_2 y_2}{m_1 + m_2}$$

$$\Rightarrow \quad Z_{CM} = \frac{m_1 z_1 + m_2 z_2}{m_1 + m_2}$$

- **More than two rigid bodies**

  (i) Centre of mass of symmetrical rigid body (like sphere, disc, cube, etc.) lies at its geometric centre.

  (ii) For two or more than two rigid bodies, we can use

$$\mathbf{r}_{CM} = \frac{m_1\mathbf{r}_1 + m_2\mathbf{r}_2}{m_1 + m_2}$$

$$\Rightarrow \quad X_{CM} = \frac{m_1x_1 + m_2x_2}{m_1 + m_2}$$

$$\Rightarrow \quad Y_{CM} = \frac{m_1y_1 + m_2y_2}{m_1 + m_2}$$

and

$$Z_{CM} = \frac{m_1z_1 + m_2z_2}{m_1 + m_2}$$

  (iii) If three dimensional rigid body has uniform density, then mass in above formulae can be replaced by volume (V).

  For example, $$\mathbf{r}_{CM} = \frac{V_1\mathbf{r}_1 + V_2\mathbf{r}_2}{V_1 + V_2}$$

  (iv) In case of two dimensional body, mass can be replaced by area (A).

  For example, $$\mathbf{r}_{CM} = \frac{A_1\mathbf{r}_1 + A_2\mathbf{r}_2}{A_1 + A_2}$$

  (v) If some portion is removed from the body, then mass can be replaced by area (A).

  For example, $$\mathbf{r}_{CM} = \frac{A_1\mathbf{r}_1 - A_2\mathbf{r}_2}{A_1 - A_2}$$ (In case of two dimensional body)

  Here, A_1 = area of whole body (without removing),
  $\mathbf{r}_1$ = position vector of its centre of mass,
  A_2 = area of removed portion and
  $\mathbf{r}_2$ = position vector of centre of mass of removed portion.

Other Formulae of Centre of Mass

- $\mathbf{F}_{CM} = \mathbf{F}_1 + \mathbf{F}_2 = \mathbf{F}_T$
- $\mathbf{p}_{CM} = \mathbf{p}_1 + \mathbf{p}_2 = \mathbf{p}_T$
- $\mathbf{v}_{CM} = \dfrac{m_1\mathbf{v}_1 + m_2\mathbf{v}_2}{m_1 + m_2} = \dfrac{\mathbf{p}_{CM}}{M_{CM}} = \dfrac{\mathbf{p}_T}{M_T}$
- $\mathbf{a}_{CM} = \dfrac{m_1\mathbf{a}_1 + m_2\mathbf{a}_2}{m_1 + m_2} = \dfrac{\mathbf{F}_{CM}}{M_{CM}} = \dfrac{\mathbf{F}_T}{M_T}$

Note *Here, T stands for total.*

Conservation of Linear Momentum

- **For a single mass or single body**
  If net force acting on the body is zero, then
  $\mathbf{p}$ = constant or $\mathbf{v}$ = constant(if mass = constant)
- **For a system of particles or system of rigid bodies**
  If net external force acting on a system of particles or system of rigid bodies is zero, then $\mathbf{p}_{CM}$ = constant or $\mathbf{v}_{CM}$ = constant.

Collision

- **Head on elastic collision** In this case, linear momentum and kinetic energy both are conserved. After solving two conservation equations, we get

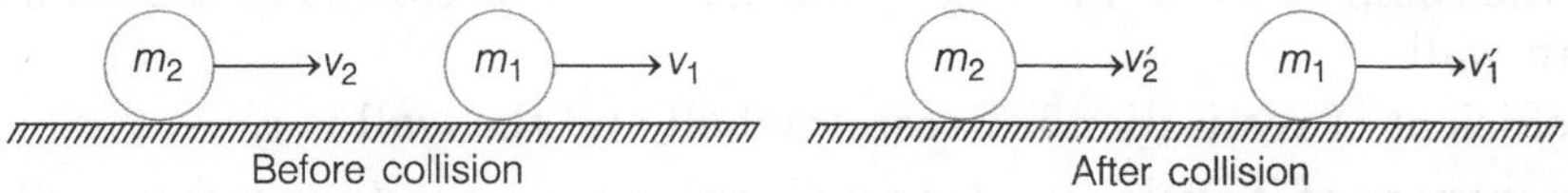

$$v_1' = \left(\frac{m_1 - m_2}{m_1 + m_2}\right) v_1 + \left(\frac{2m_2}{m_1 + m_2}\right) v_2 \quad \text{...(i)}$$

and
$$v_2' = \left(\frac{m_2 - m_1}{m_1 + m_2}\right) v_2 + \left(\frac{2m_1}{m_1 + m_2}\right) v_1 \quad \text{...(ii)}$$

In the above two formulae, following are three special cases :

(i) If $m_1 = m_2$, then $v_1' = v_2$ and $v_2' = v_1$ i.e. in case of equal masses bodies will exchange their velocities.

(ii) If $m_1 >> m_2$ and $v_1 = 0$, then $v_1' \approx 0$ and $v_2' \approx -v_2$.

(iii) If $m_2 >> m_1$ and $v_1 = 0$, then $v_1' = 2v_2$ and $v_2' \approx v_2$.

- **Head on inelastic collision** In this type of collision, only linear momentum remains constant. Suppose two unknowns velocities are v_1' and v_2'. Make following two equations to solve them.

(i) Conservation of linear momentum

$$m_1 v_1 + m_2 v_2 = m_1 v_1' + m_2 v_2'$$

(ii) Definition of coefficient of restitution (e)

$$e = \frac{|\text{ relative velocity of separation }|}{|\text{ relative velocity of approach }|} = \frac{v_1' - v_2'}{v_2 - v_1}$$

- **Perfectly inelastic collision** In perfectly inelastic collision, the colliding bodies stick together after collision and move with a common velocity given by

$$\mathbf{v} = \frac{\text{total momentum of the system}}{\text{total mass}} = \frac{\mathbf{p}_{\text{total}}}{M_{\text{total}}}$$

- **Oblique collision (both elastic and inelastic)** Resolve the velocities along common normal and common tangent directions. Now,

(i) velocity components along common tangent direction will remain unchanged.

(ii) along common normal direction, theory of head on collision (elastic as well as inelastic) can be used.

- If a body is dropped from a height h, then
  v_0 = velocity just before striking with ground = $\sqrt{2gh}$
  v_1 = velocity just after striking with ground = ev_0
  v_n = velocity just after striking n^{th} time with ground = $e^n v_0$
  h_n = height attained by the body after n^{th} collision = $\dfrac{v_n^2}{2g} = e^{2n} h$

  Here, e = coefficient of restitution.
- Suppose a ball is projected with speed u at an angle θ with horizontal. It collides at some distance with a wall parallel to y-axis as shown in figure. Let v_x and v_y be the components of its velocity along x and y-directions at the time of impact with wall.
  Coefficient of restitution between the ball and the wall is e.
  Component of its velocity along y-direction (common tangent) v_y will remain unchanged while component of its velocity along x-direction (common normal) v_x will become ev_x in opposite direction.

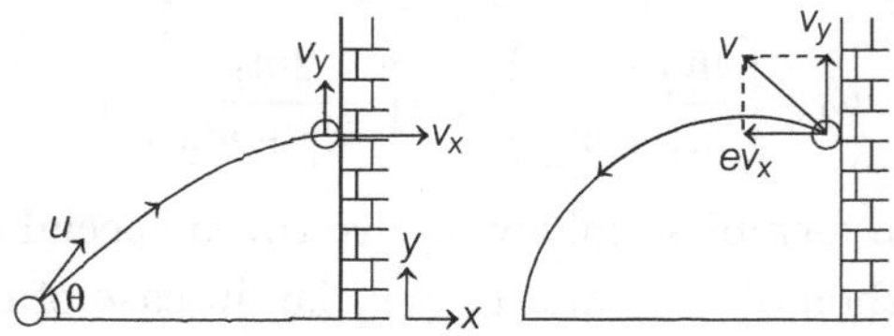

Further, since v_y does not change due to collision, the time of flight (time taken by the ball to return to the same level) and maximum height attained by the ball will remain same as it would have been in the absence of collision with the wall. Thus,

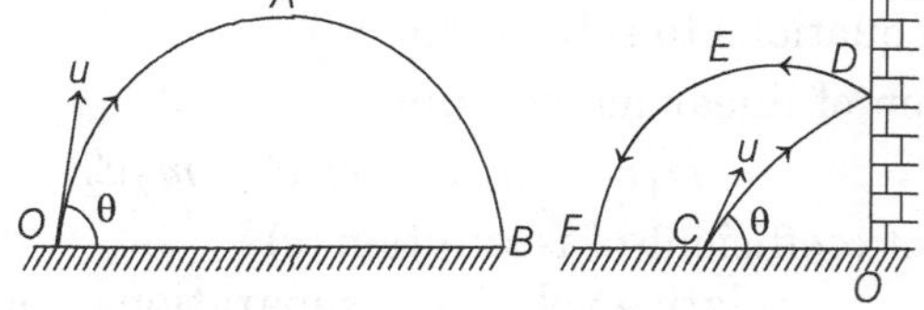

$$t_{OAB} = t_{CD} + t_{DEF} = T = \frac{2u \sin\theta}{g}$$

and $$h_A = h_E = \frac{u^2 \sin^2\theta}{2g}$$

Further, $CO + OF \le$ Range or OB

If collision is elastic, then $CO + OF = \text{Range} = \dfrac{u^2 \sin 2\theta}{g}$

and if it is inelastic, $CO + OF <$ Range

- In the projectile motion as shown in figure.

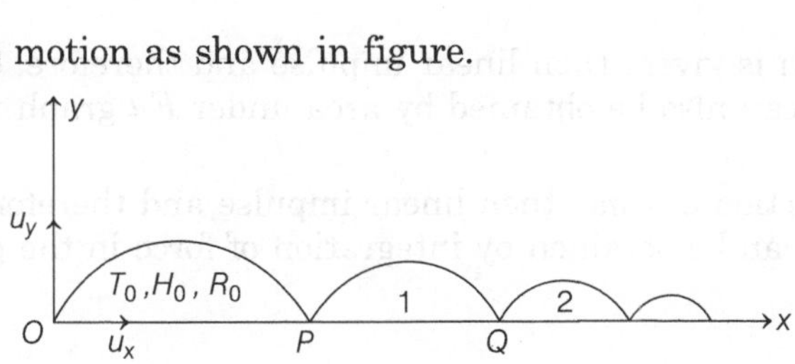

$$T = \frac{2u_y}{g} \Rightarrow T \propto u_y$$

$$\Rightarrow \quad H = \frac{u_y^2}{2g} \Rightarrow H \propto u_y^2$$

$$R = u_x T = u_x \left(\frac{2u_y}{g} \right) \Rightarrow R \propto u_x u_y$$

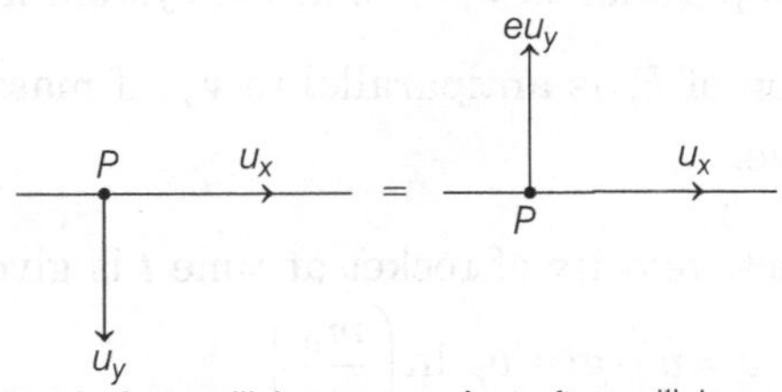

Just before collision Just after collision

As shown in above figure, vertical component of velocity just after collision becomes eu_y or e times, while horizontal component remains unchanged.

Hence, the next time T will become e times (as $T \propto u_y$), H will become e^2 times (as $H \propto u_y^2$) and R will also become e times (as $R \propto u_x u_y$).

Thus, if T_0, H_0 and R_0 are the initial values, then after first collision,

$$T_1 = eT_0, \; H_1 = e^2 H_0 \quad \text{and} \quad R_1 = eR_0$$

Similarly after n-collisions,

$$T_n = e^n T_0, \; H_n = e^{2n} H_0 \quad \text{and} \quad R_n = e^n R_0$$

Linear Impulse

- When a large force acts for a short interval of time, then product of force and time is called linear impulse. It is a vector quantity denoted by **J**. This is equal to the change in linear momentum. Thus,

  Linear impulse

$$\mathbf{J} = \mathbf{F} \cdot \Delta t = \Delta \mathbf{p} = \mathbf{p}_f - \mathbf{p}_i = m(\mathbf{v}_f - \mathbf{v}_i)$$

- In one dimensional motion, we can write

$$J = F\Delta t = \Delta p = p_f - p_i = m(v_f - v_i)$$

- In this case, we will choose a sign convention and all vector quantities are substituted with proper signs.

- If F - t graph is given, then linear impulse and therefore, change in linear momentum can also be obtained by area under F-t graph with projection along t-axis.
- If $\mathbf{F}$ is a function of time, then linear impulse and therefore, change in linear momentum can be obtained by integration of force in the given time interval.

Variable Mass

- A thrust force will act when mass of a system either increases or decreases.
  This force is given by $\quad \mathbf{F}_t = \mathbf{v}_r\left(\pm \frac{dm}{dt}\right)$

  Here, $\mathbf{v}_r$ is relative velocity of mass dm which either enters or leaves the system on which thrust force has to be applied.
- Magnitude of thrust force is given by, $F_t = \left| \mathbf{v}_r\left(\pm \frac{dm}{dt}\right) \right|$
- Direction of $\mathbf{F}_t$ is parallel to $\mathbf{v}_r$, if mass of system is increasing or $\frac{dm}{dt}$ is positive. Direction of $\mathbf{F}_t$ is antiparallel to $\mathbf{v}_r$, if mass of system is decreasing or $\frac{dm}{dt}$ is negative.

  Based on this fact, velocity of rocket at time t is given by

$$v = u - gt + v_r \ln\left(\frac{m_0}{m}\right)$$

  Here, u = initial velocity of rocket
  v_r = exhaust velocity of gases (Assumed constant)
  m_0 = initial mass of rocket (with gases)
  and m = mass of rocket at time t.

  Value of g has been assumed constant in above equation.
- If mass is just dropped from a moving body, then the mass which is dropped acquires the same velocity as that of the moving body.

  Hence, $\mathbf{v}_r = 0$ or no thrust force will act in this case although mass is decreasing.
- If two or more than two particles are in motion freely under gravity, then absolute acceleration of each particle is g (downwards). So, relative acceleration between any two particles is zero but acceleration of their centre of mass is again g (downwards).
- If a projectile explodes in air in different parts, the path of the centre of mass remains unchanged. This is because during explosion no other external force (except gravity) acts on the centre of mass. The situation is as shown in figure.

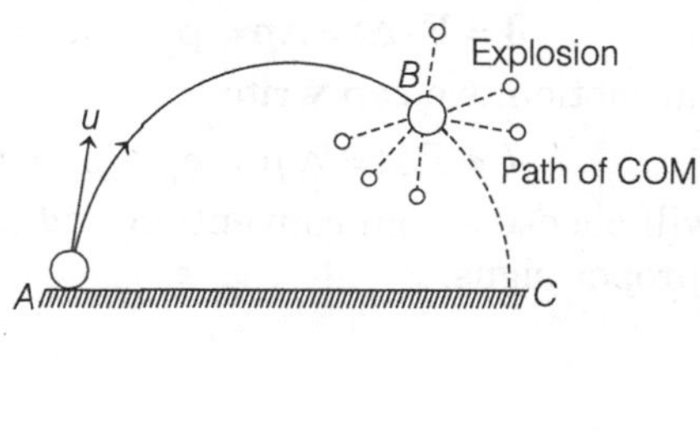

Path of COM is parabola, even though the different parts travel in different directions after explosion.

This situation continues till the first particle strikes the ground because after that force, behaviour of system of particles will change.

Centre of Mass, Frame of Reference or C-frame of Reference or Zero Momentum Frame

- A frame of reference carried by the centre of mass of an isolated system of particles (i.e. a system not subjected to any external forces) is called the centre of mass or *C*-frame of reference. In this frame of reference,
  (i) position vector of centre of mass is zero.
  (ii) velocity and hence, momentum of centre of mass is also zero.
- A liquid of density ρ is filled in a container as shown in figure. The liquid comes out from the container through a orifice of area a at a depth h below the free surface of the liquid with a velocity v. This exerts a thrust force in the container in the backward direction. This thrust force is given by

$$F_t = v_r\left(-\frac{dm}{dt}\right)$$

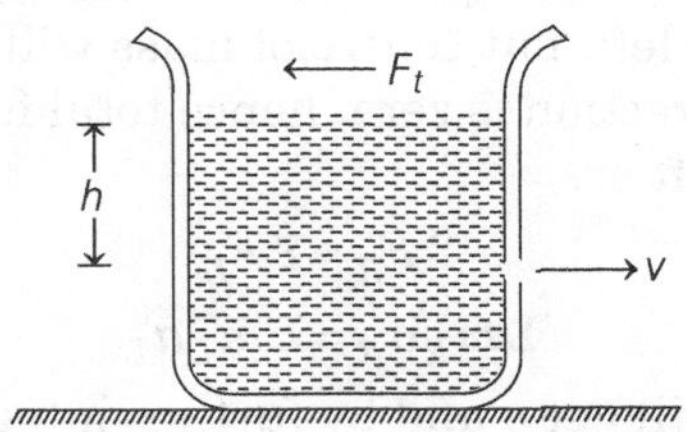

Here, $v_r = v$ (in forward direction)

and $\left(-\frac{dm}{dt}\right) = \rho a v$

As $\left(\frac{dV}{dt}\right)$ = volume of liquid flowing per second = av

$$\therefore \quad \left(-\frac{dm}{dt}\right) = \rho\left(\frac{dV}{dt}\right) = \rho a v$$

$\therefore\ F_t = v\,(\rho a v)$ or $F_t = \rho a v^2$ (in backward direction)

Here, $v = \sqrt{2gh}$.

- Suppose, a chain of mass per unit length λ begins to fall through a hole in the ceiling as shown in Fig. (a) or the end of the chain piled on the platform is lifted vertically as in Fig. (b).

  In both the cases, due to increase of mass in the portion of the chain which is moving with a velocity v at certain moment of time a thrust force acts on this part of the chain which is given by

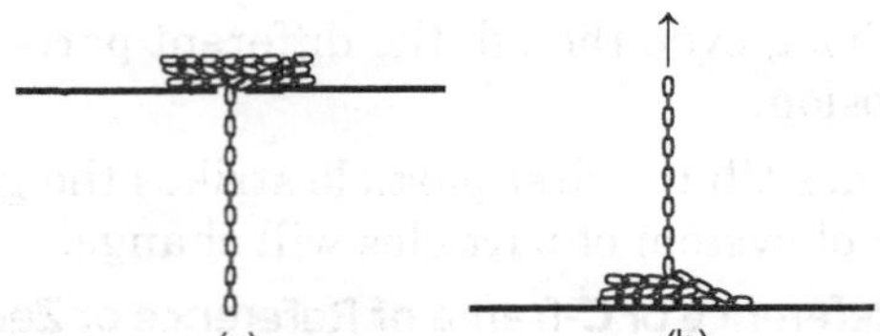

$$F_t = v_r \left(\frac{dm}{dt}\right)$$

Here, $v_r = v$ and $\frac{dm}{dt} = \lambda v$

Here, v_r is upwards in case (a) and downwards in case (b). Thus,

$$F_t = \lambda v^2$$

The direction of F_t is upwards in case (a) and downwards in case (b).

- The net force on a system in a particular direction is zero (normally in horizontal direction). This can be done by giving the horizontal ground smooth.

  Since the system is at rest initially, so in this case individual bodies can move towards right or towards left, but centre of mass will remain stationary. Further, net force in horizontal direction is zero, hence total force towards right is equal to the total force towards left or,

  $$\Sigma F_R = \Sigma F_L \quad \text{...(i)}$$

  or $$\Sigma m_R a_R = \Sigma m_L a_L \quad \text{...(ii)}$$

  Now, integrating a, we will get v and by further integrating v, we will get x.

  $\therefore$ $$\Sigma m_R v_R = \Sigma m_L v_L \quad \text{...(iii)}$$

  and $$\Sigma m_R x_R = \Sigma m_L x_L \quad \text{...(iv)}$$

- If a bomb or a projectile explodes in two or more than two parts, then it explodes due to its internal forces. Therefore, net force or net external force is zero. Hence, linear momentum of the system can be conserved just before and just after explosion.

  By this momentum conservation equation, we can find the velocity of some unknown parts. If explosion takes place in air, then during the explosion, the external force due to gravity (= weight) can be neglected, as the time of explosion is very short.

  So, impulse of this force is negligible and impulse is change in linear momentum. Hence, change in linear momentum is also negligible.

CHAPTER 08

Rotational Motion

Moment of Inertia

Thin rod	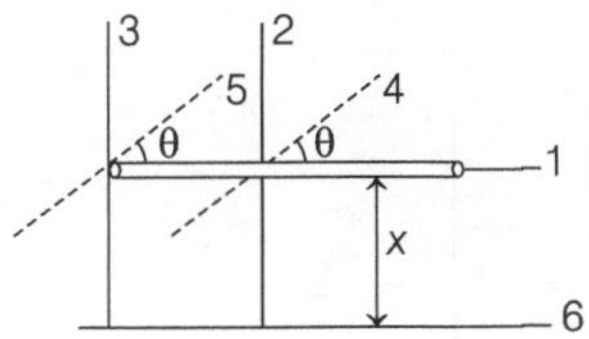	$I_1 = 0,\ I_2 = \frac{ml^2}{12}$ $I_3 = \frac{ml^2}{3},\ I_4 = \frac{ml^2}{12}\sin^2\theta$ $I_5 = \frac{ml^2}{3}\sin^2\theta,\ I_6 = mx^2$
Circular disc	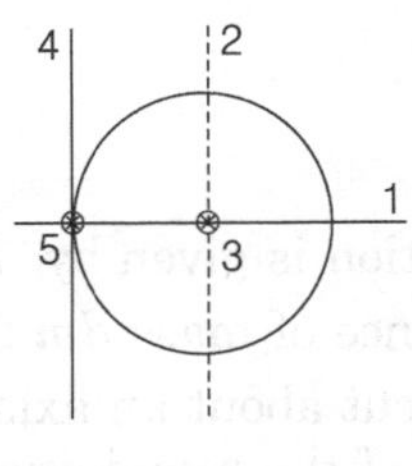	$I_1 = I_2 = \frac{mR^2}{4}$ $I_3 = I_1 + I_2 = \frac{mR^2}{2}$ $I_4 = I_2 + mR^2 = \frac{5}{4}mR^2$ $I_5 = I_3 + mR^2 = \frac{3}{2}mR^2$
Circular ring	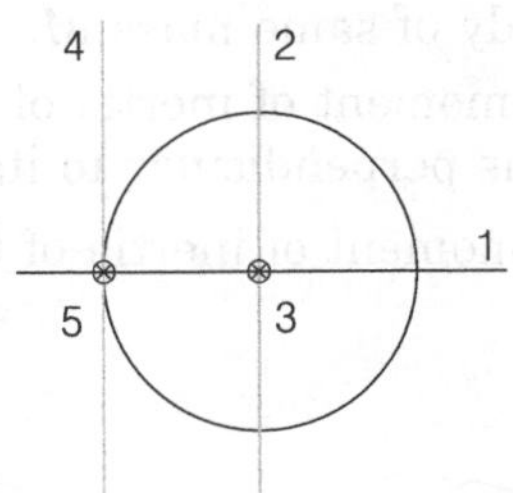	$I_1 = I_2 = \frac{mR^2}{2}$ $I_3 = I_1 + I_2 = mR^2$ $I_4 = I_2 + mR^2 = \frac{3}{2}mR^2$ $I_5 = I_3 + mR^2 = 2mR^2$
Rectangular slab	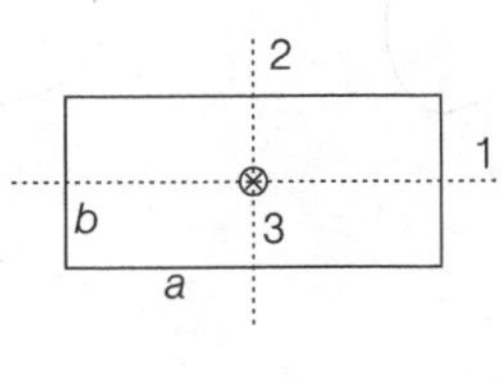	$I_1 = \frac{mb^2}{12}$ $I_2 = \frac{ma^2}{12}$ $I_3 = I_1 + I_2$ $= \frac{m}{12}(a^2 + b^2)$

Square slab	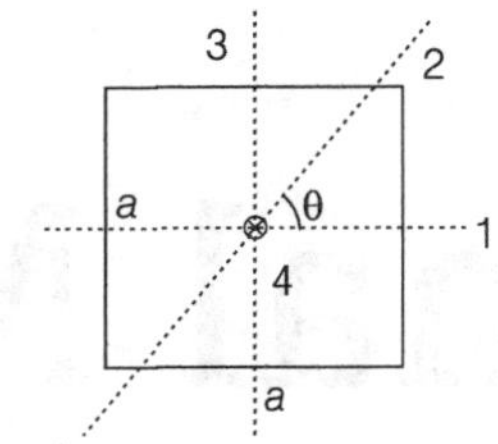	$I_1 = I_2 = I_3 = \frac{ma^2}{12}$ $I_4 = I_1 + I_3 = \frac{ma^2}{6}$
Solid sphere	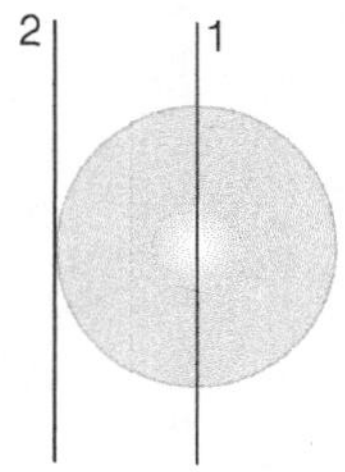	$I_1 = \frac{2}{5} mR^2$ $I_2 = I_1 + mR^2$ $= \frac{7}{5} mR^2$
Hollow sphere	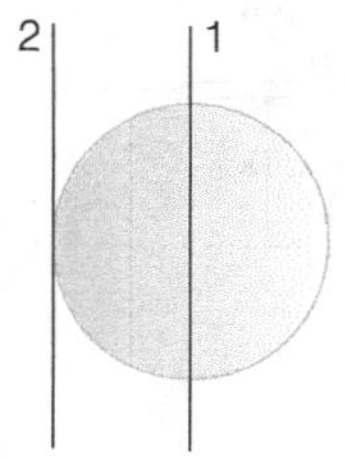	$I_1 = \frac{2}{3} mR^2$ $I_2 = I_1 + mR^2$ $= \frac{5}{3} mR^2$

- Moment of inertia by integration is given by $I = \int (dm)\, r^2$
  Here, r is perpendicular distance of mass dm from the axis.
- If a portion is symmetrically cut about an axis and mass of remaining portion is M, then moment of inertia of the remaining portion is same as the moment of inertia of the whole body of same mass M.
  For example, in Fig. (a), moment of inertia of the section shown (a part of circular disc) about an axis perpendicular to its plane and passing through point O is $\frac{1}{2} MR^2$ as the moment of inertia of the complete disc is also $\frac{1}{2} MR^2$.

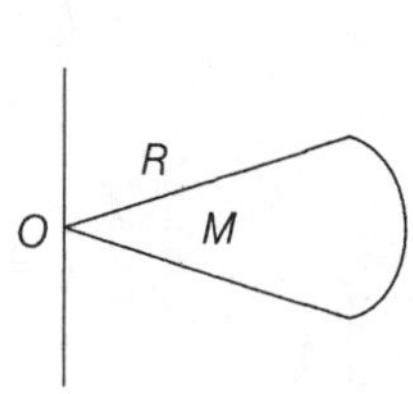

(a)

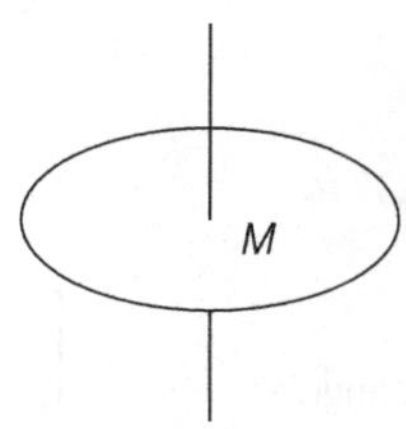

(b)

Theorems on Moment of Inertia

- There are two important theorems on moment of inertia
  (i) **Theorem of parallel axes** The moment of inertia of a body about an axis is equal to sum of moments of inertia of the body about a parallel axis, passing through its centre of mass and the product of its mass and the square of distance between the two parallel axes.

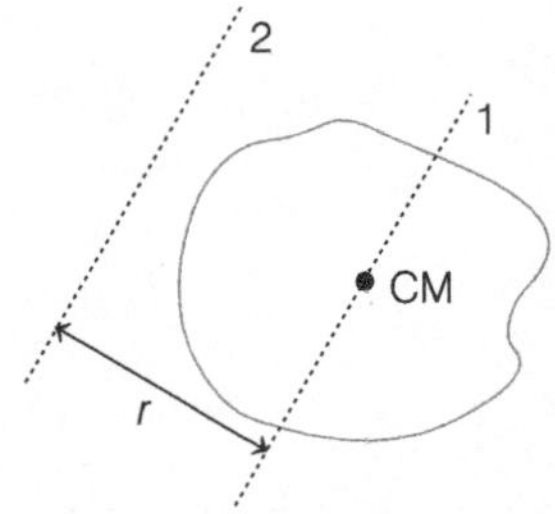

$$I = I_{CM} + mr^2$$

or

$$I_2 = I_1 + mr^2$$

  (ii) **Theorem of perpendicular axes** This theorem is applicable only for a two dimensional body of negligible thickness. If x and y are two perpendicular axes lying in the plane of body and z is the axis perpendicular to plane of body and passing through point of intersection of x and y, then

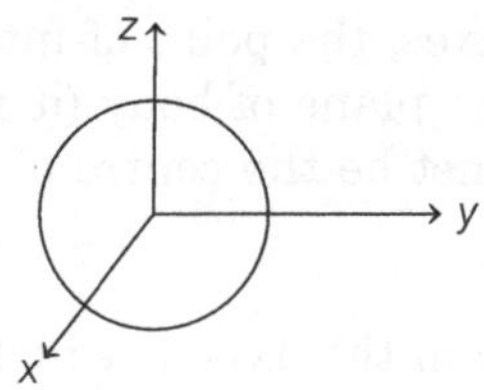

$$I_z = I_x + I_y$$

- From the first theorem, we can see that among several parallel axes, moment of inertia is least about an axis which passes through centre of mass. For example I_2 is least among I_1, I_2 and I_3. Similarly, I_5 is least among I_4, I_5 and I_6.

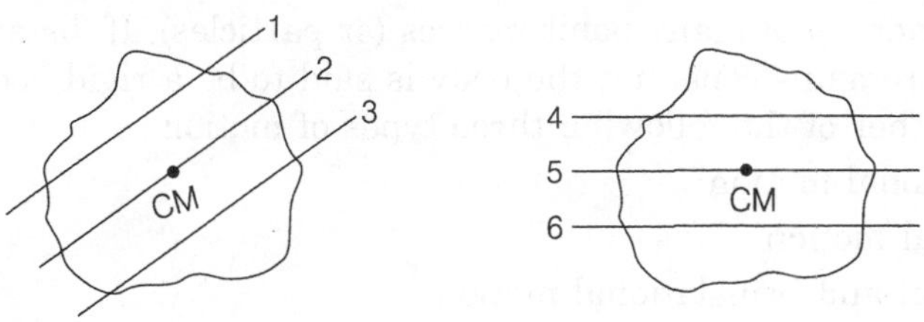

- From $I = I_{CM} + mr^2$, we can see that if we draw a graph between I and r (among too many parallel axes), then it is a parabola as shown in figure below.

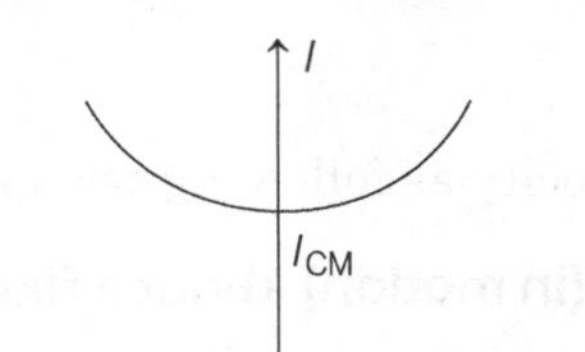

- In theorem of parallel axes, moment of inertia from centre of mass axis cannot be found independently without considering I_{CM}.

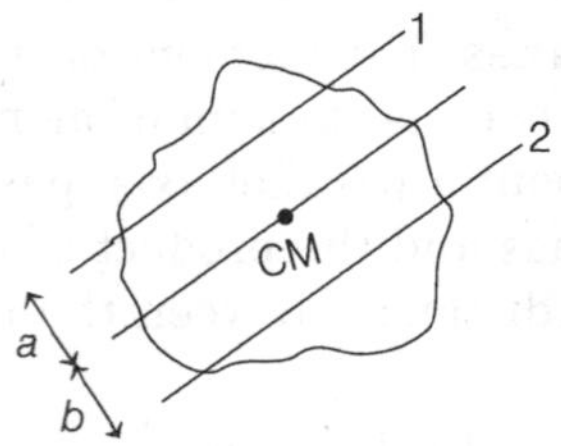

For example, if I_1 is known and we wish to find I_2, then we cannot find I_2 directly as,

$$I_2 = I_1 + m(a+b)^2$$

The correct approach is as under, first you find I_{CM} and then I_2.

$$I_{CM} = I_1 - ma^2 \quad \text{and now,} \quad I_2 = I_{CM} + mb^2$$

- In theorem of perpendicular axes, the point of intersection of the three axes (x, y and z) may be any point on the plane of body (it may even lie outside the body also). This point may or may not be the centre of mass of the body.

Radius of Gyration (K)

It is an imaginary distance from the axis at which whole mass of the rigid body, if kept as a point mass, moment of inertia remains unchanged.

Thus, $$I = mK^2 \quad \text{or} \quad K = \sqrt{\frac{I}{m}}$$

Three Types of Motion of a Rigid Body

- A rigid body is made up of many point masses (or particles). If distance between any two particles remains constant, the body is said to be a rigid body. A rigid body may have either of the following three types of motion
  - (i) only translational motion
  - (ii) only rotational motion
  - (iii) both rotational and translational motion.
- Only in case of translational motion displacement, velocity and acceleration of all particles are same. In rotational or rotational plus translational motion, different particles have different displacements, velocities and accelerations.

Angular Velocity

We may define angular velocity as following two types:

1. Angular velocity of a particle (in motion) about a fixed point

- At a given instant , a particle P has velocity $\mathbf{v}$.

  It has position vector $\mathbf{r}$ with respect to a fixed point O as shown in figure.

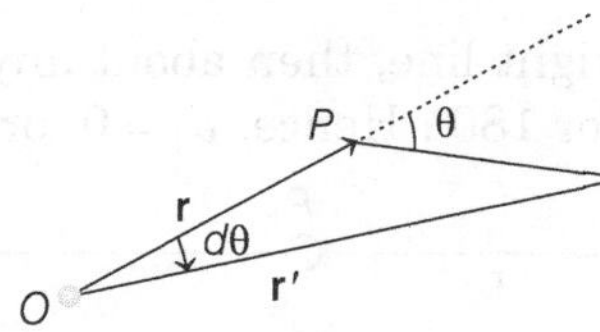

After some time, position vector has become **r′**. We can see two changes in its position vector.

First, its magnitude $|\mathbf{r}|$ has changed, second its direction has changed or we can say, its position vector has been rotated.

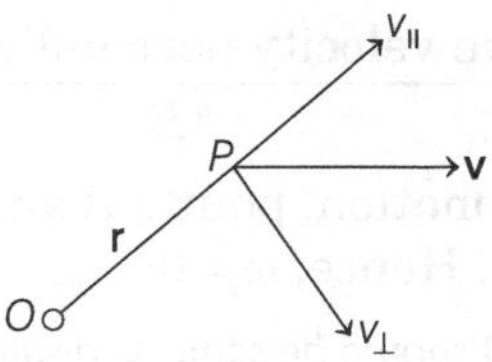

If we resolve **v** along **r** and perpendicular to **r**, then the two components $v_{\parallel}$ and $v_{\perp}$ have the following meanings

$$v_{\parallel} = v\cos\theta = \frac{\mathbf{v}\cdot\mathbf{r}}{r} = \text{component of } \mathbf{v} \text{ along } \mathbf{r}$$

Here, θ = angle between **r** and **v**

$$= \frac{d|\mathbf{r}|}{dt}.$$

= rate by which magnitude of **r** changes

= rate by which distance of P from O changes.

$$\frac{v_{\perp}}{|\mathbf{r}|} = \frac{v_{\perp}}{r} = \frac{v_{\perp}}{OP} = \omega$$

= angular velocity of particle P about point O at this instant

= rate by which **r** rotates

- If θ is acute, $\cos\theta$ or $v_{\parallel}$ is positive, i.e. distance of P from O is increasing.

  If θ is obtuse, $\cos\theta$ or $v_{\parallel}$ is negative, i.e. distance of P from O is decreasing.

  If θ is 90°, then $\cos\theta$ or $v_{\parallel}$ is zero or distance of P from O is constant.

  For example, when particle rotates in a circle, then with respect to centre, θ is always 90°.

  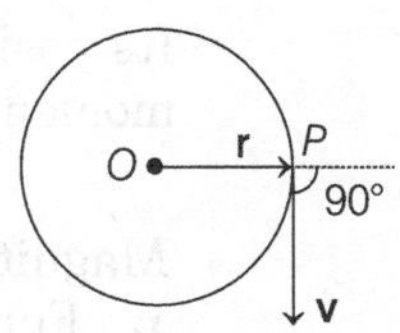

  This is the reason, why distance of particle from centre always remains constant. With respect to any other point, θ is sometimes acute and sometimes obtuse. Therefore, distance sometimes increases and sometimes decreases.
- $\mathbf{v}_{\perp} = \boldsymbol{\omega}\times\mathbf{r}$, i.e. perpendicular component of velocity in vector form is the cross product of ω and **r**.
- Direction of ω is given by right hand screw law.

- If a particle moves in a straight line, then about any point lying on this line, angle between **r** and **v** is 0° or 180°. Hence, $v_{\perp} = 0$ or $\omega = 0$.

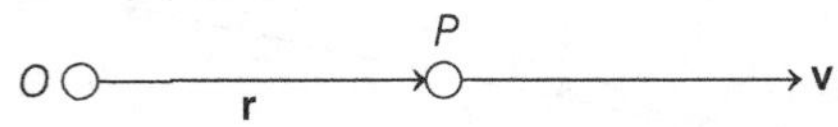

2. Angular velocity of a rigid body

Take any two points A and B on the rigid body.

The velocities of A and B may or may not be same. In this case, angular velocity of rigid body is defined as

$$\omega = \frac{\text{relative velocity perpendicular to } AB}{AB}$$

In case of pure translational motion, points A and B will have same velocities. Therefore, numerator is zero. Hence, $\omega = 0$.

Note *Components of velocities along AB should be same as distance AB is constant for a rigid body.*

Torque

$\tau = \mathbf{r} \times \mathbf{F}$. Here **r** is position vector of point of application of force (say point P) with respect to the point (say O) about which torque is required. Therefore,

$$\mathbf{r} = \mathbf{r}_P - \mathbf{r}_O$$

Angular Momentum (L)

Angular momentum can be defined in following three ways

1. Angular momentum of a particle (*A*) in motion, about a fixed point (*O*)

- Suppose a particle A has a linear momentum $\mathbf{p} = m\mathbf{v}$ as shown in the figure.

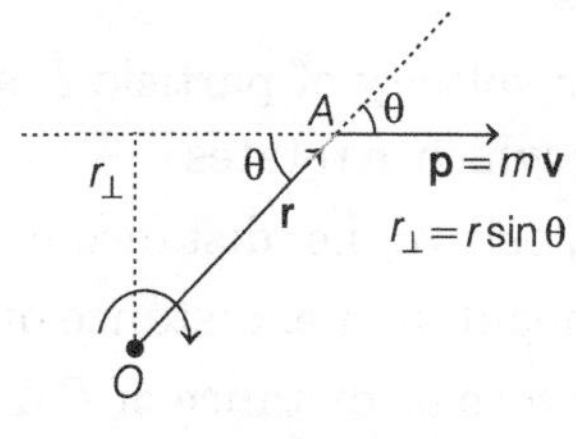

Its position vector about a fixed point O at this instant is **r**. Then, angular momentum of particle A about point O will be

$$\mathbf{L} = \mathbf{r} \times \mathbf{p} = \mathbf{r} \times (m\mathbf{v}) = m\,(\mathbf{r} \times \mathbf{v})$$

- Magnitude of **L** is $L = mvr \sin\theta = mvr_{\perp}$, where θ is the angle between **r** and **p**. Further, $r_{\perp} = r \sin\theta$ is the perpendicular distance on line of action of **p** (or **v**) from point O. Direction of **L** will be given by right hand screw law. In the shown figure, direction of **L** is perpendicular to paper inwards.
- If line of action of velocity passes through the point O, then $r_{\perp} = 0$. Therefore, angular momentum is zero.

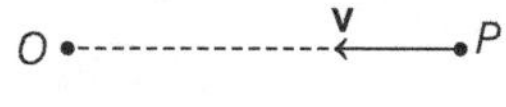

- If a particle rotates in a circle, then $r_\perp$ from centre is always equal to R (= radius of circle) and θ between **r** and **v** (or **p**) is always 90°, therefore angular momentum about centre is $L = mvR$

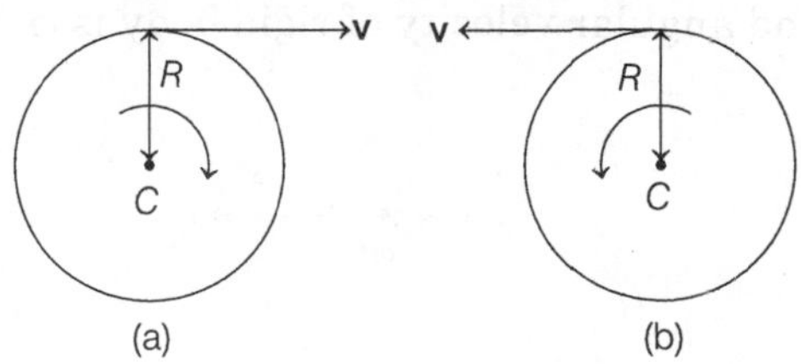

In Fig. (a), direction of angular momentum is perpendicular to paper inwards or ⊗ and in Fig. (b) outwards or ⊙.

- If a particle is moving with constant velocity (speed and direction both are constant), then angular momentum about any point always remains constant. But this constant value will be different about different points.

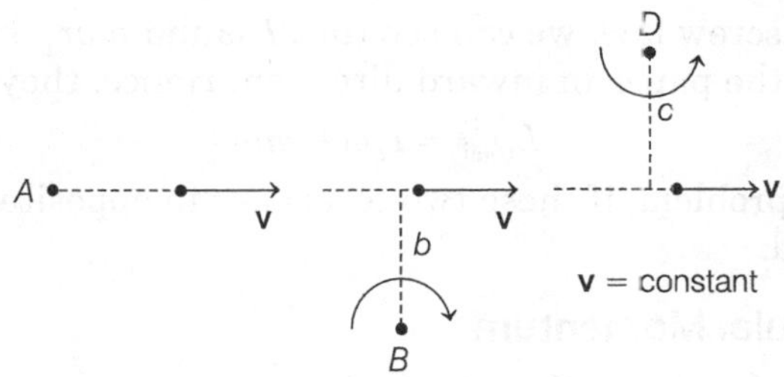

In the figure shown above,

$$L_A = 0 \quad \text{as} \quad r_\perp = 0$$

$L_B = mvb$ (= constant), perpendicular to paper inwards

and $L_D = mvc$ (= constant), perpendicular to paper outwards.

2. Angular Momentum of a Rigid Body in Pure Rotation About Axis of Rotation

If a rigid body is in pure rotation about a fixed axis, then angular momentum of rigid body about this axis will be given by $L = I\omega$

This is actually component of total angular momentum about axis of rotation. Direction of this component is again given by right hand screw law.

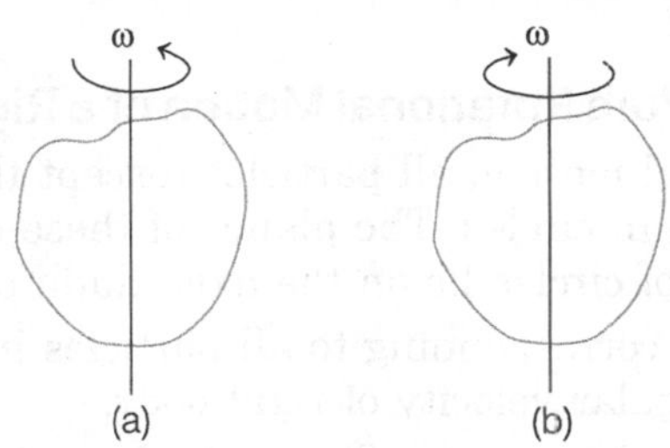

In Fig. (a), this is along the axis in upward direction.

In Fig. (b), this is along the axis in downward direction.

3. Angular Momentum of a Rigid Body in Rotation Plus Translation About a General Axis

Suppose a rigid body is in rotational and translational motion and velocity of its centre of mass is v and angular velocity of rigid body is ω.

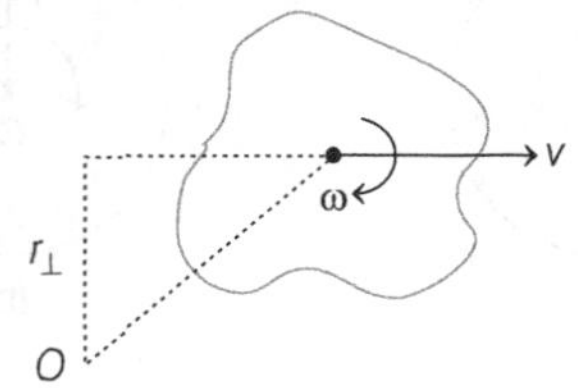

We want to find angular momentum of rigid body about an axis passing from a general point O and perpendicular to plane of paper.

This will contain two terms :

(a) $I_c\omega$ (b) $mv_c r_\perp = mvr_\perp$

From right hand screw law, we can see that $I_c\omega$ and $mvr_\perp$ both terms are perpendicular to the paper in inward direction. Hence, they are added.

or $$L_{\text{total}} = I_c\omega + mvr_\perp$$

In some another problem, if these two terms are in opposite directions, then they will be subtracted.

Conservation of Angular Momentum

According to law of conservation of angular momentum,

$$\tau = \frac{d\mathbf{L}}{dt}$$

If $\tau = 0$, then $\mathbf{L}$ = constant.

In pure rotational motion, $L = I\omega$ = constant

or $$I_1\omega_1 = I_2\omega_2$$

or $$\omega \propto \frac{1}{I}$$

If mass moves away from the axis of rotation, then I increases, therefore ω decreases or time period of rotation $T\left(= \frac{2\pi}{\omega}\right)$ increases.

Equations of Motion of Pure Rotational Motion of a Rigid Body about a Fixed Axis

- In pure rotational motion, all particles (except those lying on the axis) of the rigid body rotate in circles. The planes of these circles are mutually parallel and the centres of circles lie on the axis. Radii of different circles are different.
- Angular velocity corresponding to all particles in circular motion is same. This is also called angular velocity of rigid body.
- Plane of every circle is perpendicular to the axis of rotation.

- Velocity of any particle P is $v = r\,\omega$, tangential to its own circle. Since ω is same for all particles, so we can say that $v \propto r$.

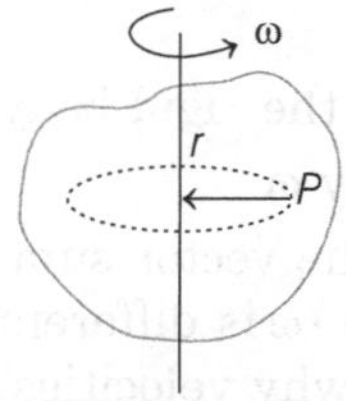

- Acceleration of particle P will have two components a_n and a_t as it is rotating in a circle. Here,

$$a_n = r\omega^2 \text{ or } \frac{v^2}{r} \quad \text{(Towards centre)}$$

$$a_t = r\alpha = r \cdot \frac{d\omega}{dt} \quad \text{(Tangential to the circle)}$$

$$\therefore \quad a = \sqrt{a_n^2 + a_t^2}$$

- If angular velocity ω is constant, then angular acceleration

$$\alpha = \frac{d\omega}{dt} = 0$$

and $\theta = \omega t$

- If α is constant, then

$$\omega = \omega_0 + \alpha t$$

$$\theta = \omega_0 t + \frac{1}{2}\alpha t^2$$

and $\omega^2 = \omega_0^2 + 2\alpha\theta$

Here, ω is angular velocity at time t and ω_0 is initial angular velocity.
If α is not constant, we will have to go for differentiation or integration.
The basic equations of differentiation or integration are

$$\omega = \frac{d\theta}{dt}, \alpha = \frac{d\omega}{dt} = \omega \cdot \frac{d\omega}{d\theta} \quad \text{(Equations of differentiation)}$$

$$\int d\theta = \int \omega dt, \int d\omega = \int \alpha dt, \int \omega\, d\omega = \int \alpha \cdot d\theta$$

(Equations of integration)

- Number of rotations made by rigid body,

$$N = \frac{\text{angle rotated}}{2\pi} = \frac{\theta}{2\pi}$$

Rotational Plus Translational Motion of a Rigid Body

- A complex motion of rotation plus translation can be simplified by considering,
  (i) the translational motion with the velocity of centre of mass of the rigid body and
  (ii) rotation about centre of mass with angular velocity ω of rigid body.

- In this type of motion, velocities of different points of the rigid body are different. To find velocity of a general point (say *P*), we require following two quantities
  (i) velocity of centre of mass of the rigid body v
  (ii) angular velocity of rigid body ω
  Now, velocity of point *P* is the vector sum of two terms; v and $r\omega$. Here, v is common for all points, while $r\omega$ is different for different points, as r is different. That's the reason why velocities of different points are different.
  In the figure shown,

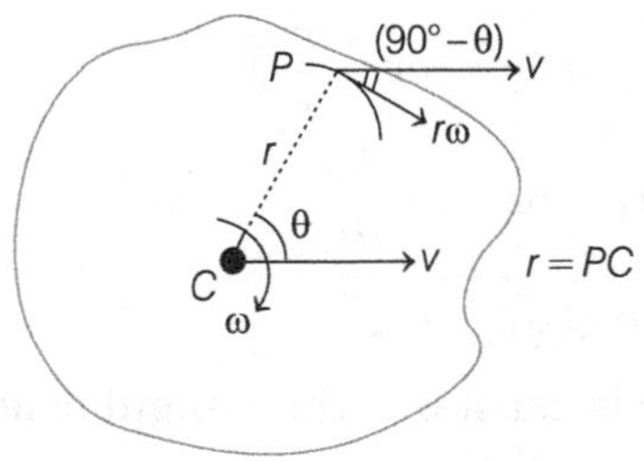

$$v_P = \sqrt{v^2 + (r\omega)^2 + 2(v)(r\omega)\cos(90° - \theta)}$$
$$= \sqrt{v^2 + r^2\omega^2 + 2vr\omega\sin\theta}$$

- To find acceleration of point *P* ,we will be required following three quantities :
  (i) acceleration of centre of mass of the rigid body a.
  (ii) angular velocity of rigid body ω and
  (iii) angular acceleration $\left(\alpha = \dfrac{d\omega}{dt}\right)$ of the rigid body.
  Now, acceleration of point *P* is the vector sum of three terms;
  a, $a_n = r\omega^2$ (acting towards centre *O*) and $a_t = r\alpha$ (acting tangentially).

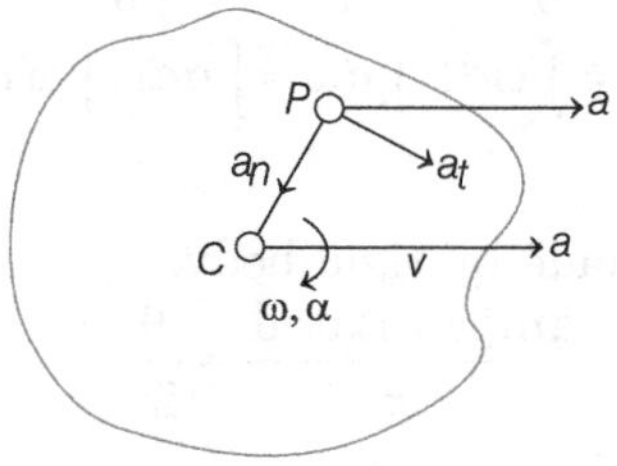

Again, a is common for all points, while a_n and a_t are different. Hence, net acceleration of different points is different.

Pure Rolling

Pure rolling (also called rolling without slipping) may be of two types

1. Uniform Pure Rolling

- In which v and ω remain constant. Condition of pure rolling (with proper sense) is $v = R\omega$. In this case, bottommost point of the body is at rest. It has no slipping with its contact point on ground because ground point is also at rest.

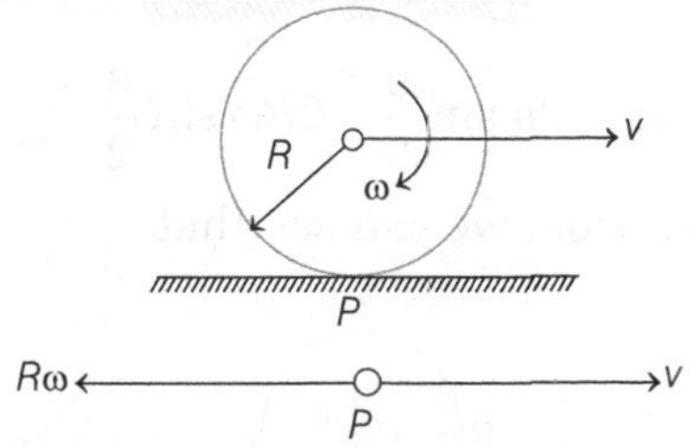

- If $v > R\omega$, then net velocity of point P is in the direction of v (in the direction of motion of body). This is called forward slipping.
- If $v < R\omega$, then net velocity of point P is in opposite direction of v. This is called backward slipping.

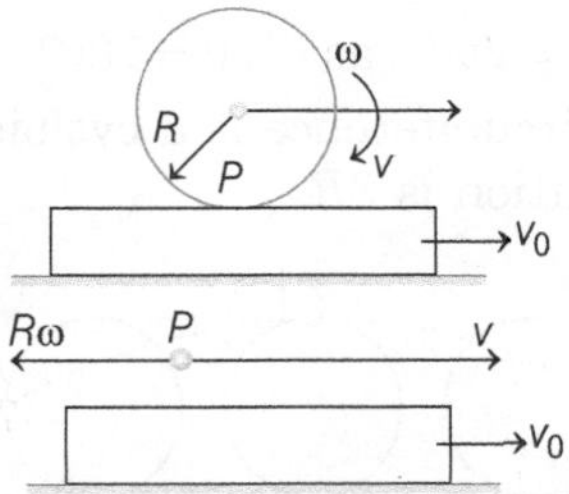

- If a body is rolling over a plank, condition for no slipping between body and plank is
$$v - R\omega = v_0$$
- In case of pure rolling on a stationary horizontal ground (when $v = R\omega$), following points are important to note
  (i) Distance moved by the centre of mass of the rigid body in one full rotation is $2\pi R$.

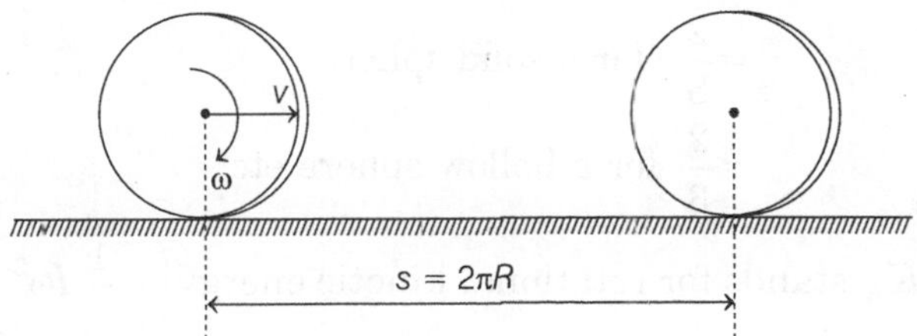

This is because $s = v \cdot T = (\omega R)\left(\frac{2\pi}{\omega}\right) = 2\pi R$

In forward slipping $s > 2\pi R$ (As $v > \omega R$)
and in backward slipping $s < 2\pi R$ (As $v < \omega R$)

(ii) The speed of a point on the circumference of the body at the instant shown in figure is $2v \sin\frac{\theta}{2}$ or $2R\omega \sin\frac{\theta}{2}$, i.e.

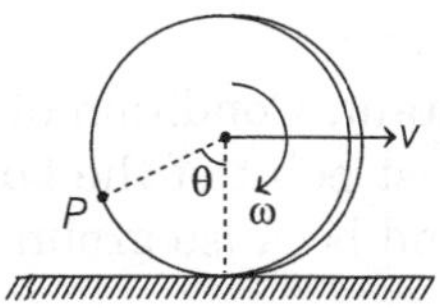

$$|\mathbf{v}_P| = v_p = 2v\sin\frac{\theta}{2} = 2R\omega\sin\frac{\theta}{2}$$

(iii) From the above expression, we can see that

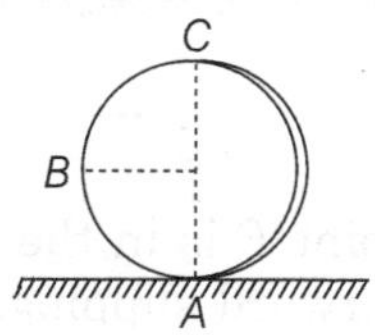

$$v_A = 0 \quad \text{as} \quad \theta = 0°$$
$$v_B = \sqrt{2}v \quad \text{as} \quad \theta = 90°$$
and
$$v_C = 2v \quad \text{as} \quad \theta = 180°$$

(iv) The path of a point on circumference is a cycloid and the distance moved by this point in one full rotation is $8R$.

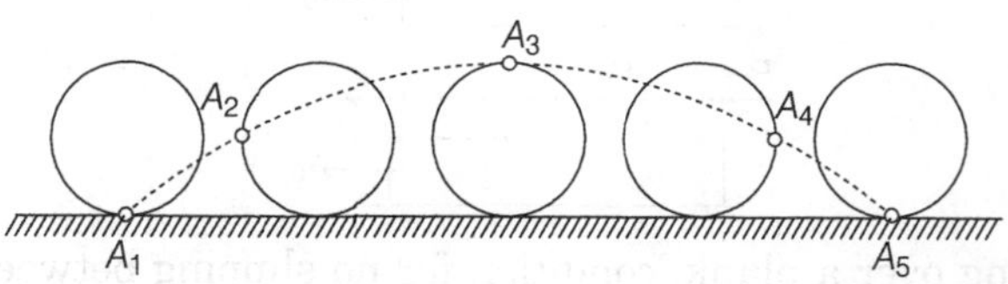

In the figure, the dotted line is a cycloid and the distance $A_1 A_2 \ldots A_5$ is $8R$.

(v) $\frac{K_R}{K_T} = 1$ for a ring

$= \frac{1}{2}$ for a disc

$= \frac{2}{5}$ for a solid sphere

$= \frac{2}{3}$ for a hollow sphere etc.

Here, K_R stands for rotational kinetic energy $\left(= \frac{1}{2} I\omega^2\right)$

and K_T for translational kinetic energy $\left(= \frac{1}{2} mv^2\right)$.

2. Accelerated Pure Rolling

If v and ω are not constant, then $a = R\alpha$ is an additional condition for pure rolling on horizontal ground, which takes place in the presence of some external forces. Here, friction plays very important role. Magnitude and direction of friction is so adjusted that equation $a = R\alpha$ is satisfied. If friction is insufficient for satisfying the equation $a = R\alpha$, slipping (either forward or backward) will occur and kinetic friction will act.

Motion of a Body on Rough Inclined Surface

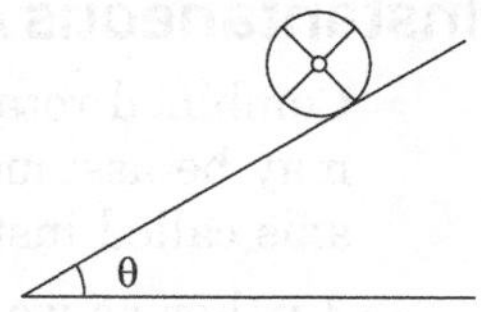

- Minimum value of coefficient of friction required for pure rolling,

$$\mu_{min} = \frac{\tan\theta}{1 + \frac{mR^2}{I}}$$

- If $\mu = 0$, body will slip downwards (only translational motion) with an acceleration, $a_1 = g\sin\theta$
- If $\mu > \mu_{min}$, body will roll (rotation + translation both) downwards without slipping with an acceleration,

$$a_2 = \frac{g\sin\theta}{1 + \frac{I}{mR^2}}$$

- In the above case (when $\mu > \mu_{min}$), force of friction will act upwards. Magnitude of this force is,

$$f = \frac{mg\sin\theta}{1 + \frac{mR^2}{I}}$$

- If $\mu < \mu_{min}$, body will roll downwards with forward slipping. Maximum friction $(= \mu mg\cos\theta)$ will act in this case. The acceleration of body is

$$a_3 = g\sin\theta - \mu g\cos\theta$$

Note *a_1 and a_3 are independent of I. But a_2 is dependent of I.*

Angular Impulse

Linear impulse when multiplied by perpendicular distance (from centre of mass) gives angular impulse. Angular impulse is equal to change in angular momentum.

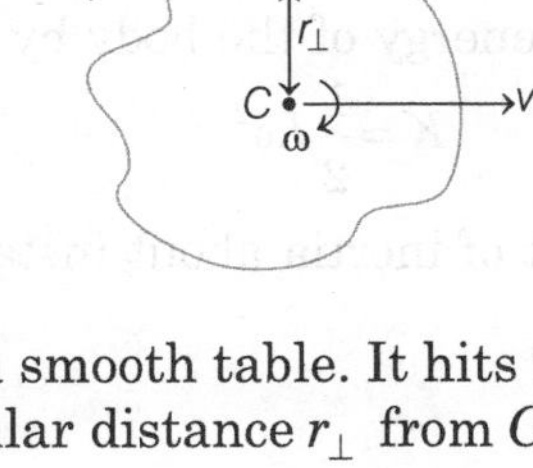

A rigid body is kept over a smooth table. It hits the table at point P by a linear impulse J at a perpendicular distance $r_\perp$ from C as shown.

Since, it hits the table at some perpendicular distance from C, so its motion is rotational plus translational. Velocity of centre of mass will be given by

$$v = \frac{J}{m} \qquad \text{(As } J = mv\text{)}$$

Angular velocity of rigid body is, $\omega = \dfrac{J \times r_\perp}{I}$

(as $J \times r_\perp$ = angular impulse = change in angular momentum = $I\omega$).

Instantaneous Axis of Rotation

- Combined rotational and translational motion of a rigid body can be simplified and may be assumed to be in pure rotational motion (with ω of rigid body) about an axis called instantaneous axis of rotation.
- Further as we know that in pure rotational motion, points lying on the axis of rotation are at rest. Therefore, we can say that, instantaneous axis of rotation passes through those points which are at rest.

  For example, in pure rolling over ground, instantaneous axis of rotation (IAOR) passes through the bottommost point, as it is a point of zero velocity.

  Thus, the combined motion of rotation and translation can be assumed to be pure rotational motion about bottommost point with same angular speed ω

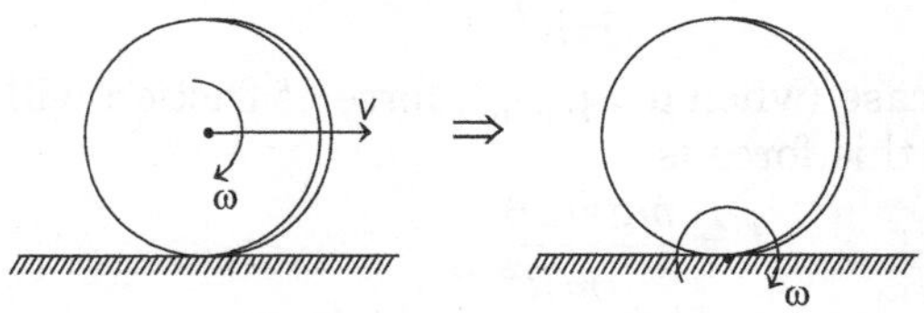

- Now, there are following uses of the concept of instantaneous axis of rotation:

  (i) Velocity of any point P can obtained by a single term

$$v = r\omega$$

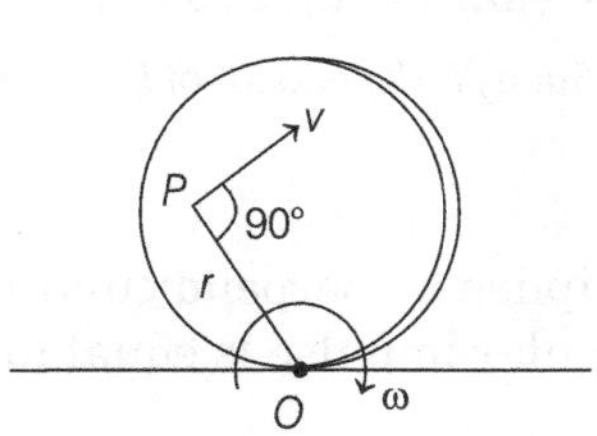

  (ii) We can find total kinetic energy of the body by a single term,

$$K = \frac{1}{2} I\omega^2$$

  But here, I is the moment of inertia about instantaneous axis of rotation.

(iii) Given the lines of action of two non-parallel velocities.

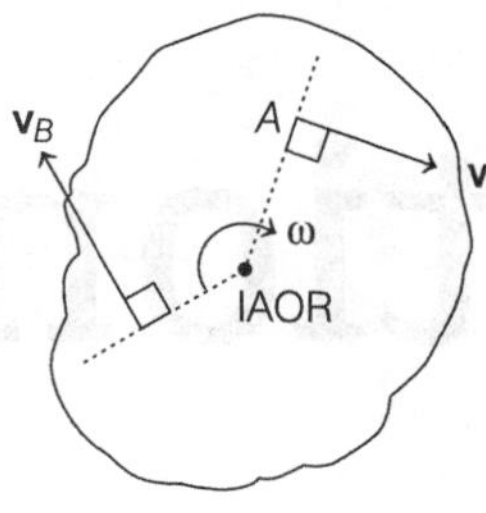

Consider the body shown in figure where the line of action of the velocities $\mathbf{v}_A$ and $\mathbf{v}_B$ are known. Draw perpendiculars at A and B to these lines of action. The point of intersection of these perpendiculars as shown locates the IAOR at the instant considered.

Principle of Moments

A lever is a light rod pivoted at point O. This point is called the fulcrum.

Seesaw on children's playground is an example of a lever.

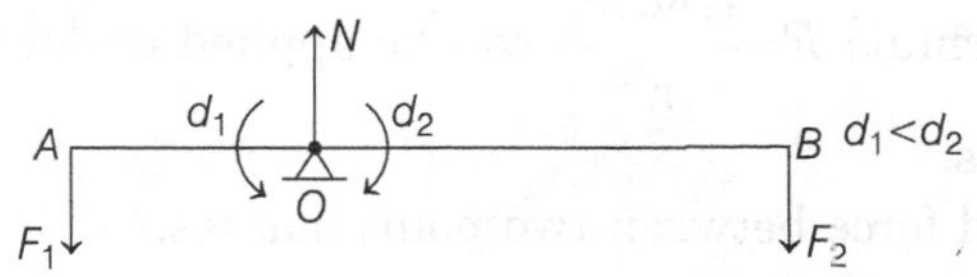

Let N be the normal reaction on rod at point O. For translational equilibrium of rod,

$$N = F_1 + F_2 \qquad \text{...(i)}$$

where, F_1 and F_2 are two parallel forces acting on rod at end points A and B.

For rotational equilibrium of rod,

Anti-clockwise moment of F_1 about O = Clockwise moment of F_2 about O

$$\therefore \quad F_1 d_1 = F_2 d_2 \qquad \text{...(ii)}$$

F_1 is normally the weight to be lifted. Distance d_1 is called load arm. Force F_2 is the effort applied to lift the load. Distance d_2 is called the effort arm.

Eq. (ii) expresses the principle of moments for a lever. The ratio $\frac{F_1}{F_2}$ is called the Mechanical Advantage (MA).

From Eq. (ii), we get

$$\text{MA} = \frac{F_1}{F_2} = \frac{d_2}{d_1} \qquad \text{...(iii)}$$

If $d_2 > d_1$, MA > 1 and $F_2 < F_1$, this means a small effort can be used to lift a large load.

CHAPTER 09

Gravitation

Gravitational Force between Two Point Masses

- It can be given as

$$F = G\frac{m_1 m_2}{r^2}$$

- Direct formula $F = \frac{Gm_1 m_2}{r^2}$ can be applied under following three conditions:

  (i) To find force between two point masses.

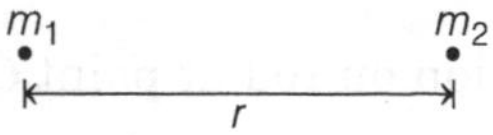

  (ii) To find force between two spherical bodies.

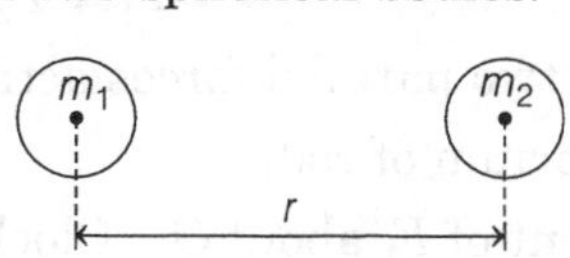

  (iii) To find force between a spherical body and a point mass.

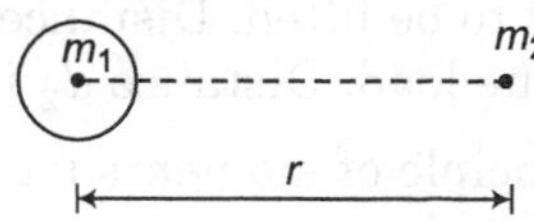

- To find force between a point mass and a rod, integration is required. In this case, we cannot assume whole mass of the rod at its centre to find force between them. Thus,

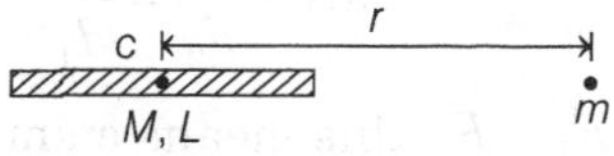

$$F \neq \frac{GMm}{r^2}$$

Acceleration Due to Gravity

- On the surface of earth, $g = \dfrac{GM}{R^2} = 9.81\ \text{m/s}^2$
- At height h from the surface of earth,

$$g' = \frac{g}{\left(1 + \frac{h}{R}\right)^2} \approx g\left(1 - \frac{2h}{R}\right), \text{ if } h << R$$

- At depth d from the surface of earth,

$$g' = g\left(1 - \frac{d}{R}\right)$$

$$g' = 0, \text{ if } d = R, \text{ i.e. at centre of earth}$$

- Effect of rotation of earth at latitude ϕ,

$$g' = g - R\omega^2 \cos^2 \phi$$

At equator $\phi = 0$, $g' = g - R\omega^2$ = minimum value
At poles, $\phi = 90°$, $g' = g$ = maximum value.
At equator, effect of rotation of earth is maximum and value of g is minimum.
At pole, effect of rotation of earth is zero and value of g is maximum.

Field Strength

- Gravitational field strength at a point in gravitational field is defined as

$$\mathbf{E} = \frac{\mathbf{F}}{m} = \text{gravitational force per unit mass}$$

- **Due to a point mass**

$$E = \frac{GM}{r^2} \quad \text{(Towards the mass)}$$

or $E \propto \dfrac{1}{r^2}$

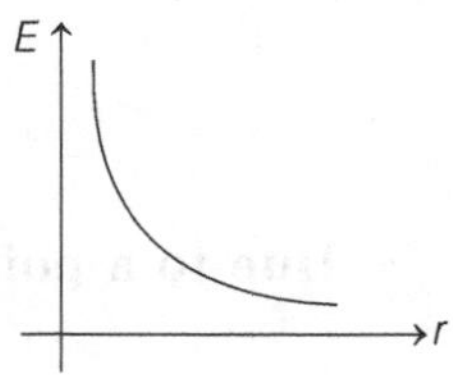

- **Due to a solid sphere**

Inside points,

$$E_i = \frac{GM}{R^3} r$$

At $r = 0$, i.e. at centre $E = 0$
At $r = R$, i.e. on surface $E = \dfrac{GM}{R^2}$

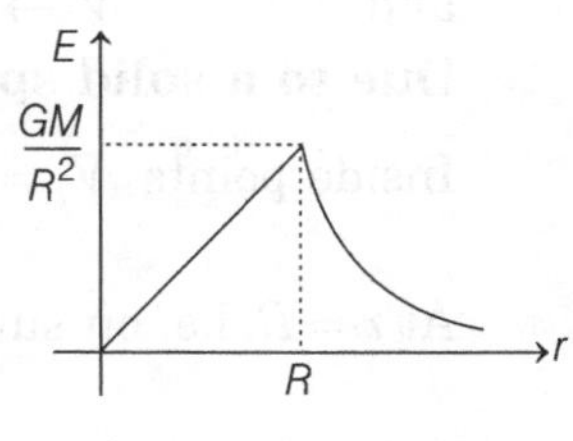

Outside points, $E_o = \dfrac{GM}{r^2}$ or $E_o \propto \dfrac{1}{r^2}$

At $r = R$, i.e. on surface $E = \dfrac{GM}{R^2}$

As $r \to \infty, E \to 0$

- **Due to a spherical shell**

Inside points, $E_i = 0$

Outside points, $E_o = \frac{GM}{r^2}$

Just outside the surface, $E = \frac{GM}{R^2}$

On the surface, E - r graph is discontinuous.

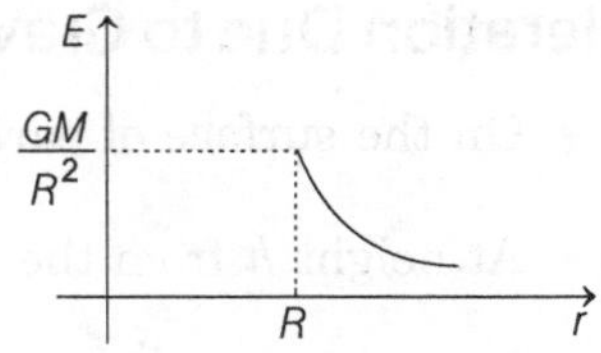

- **On the axis of a ring**

$$E_x = \frac{GMx}{(R^2 + x^2)^{3/2}}$$

At $x = 0, E = 0$ i.e. at centre $E = 0$

If $x >> R, E \approx \frac{GM}{x^2}$

i.e. ring behaves as a point mass.

As $x \to \infty, E \to 0$

$$E_{max} = \frac{2\,GM}{3\sqrt{3}R^2} \text{ at } x = \frac{R}{\sqrt{2}}$$

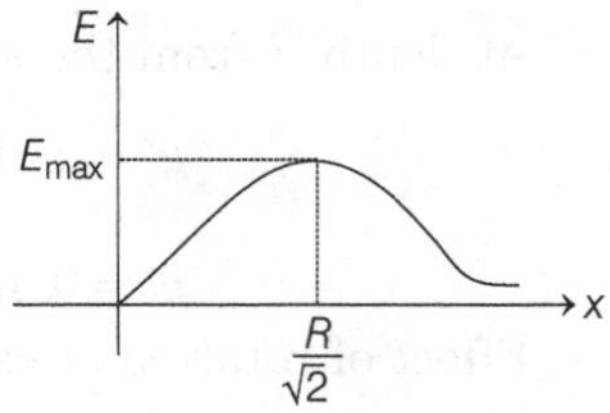

Gravitational Potential

- Gravitational potential at a point in a gravitational field is defined as the negative of work done by gravitational force in moving a unit mass from infinity to that point. Thus,

$$V_P = -\frac{W_{\infty \to P}}{m}$$

- **Due to a point mass**

$$V = -\frac{Gm}{r}$$

$V \to -\infty$ as $r \to 0$

and $V \to 0$ as $r \to \infty$

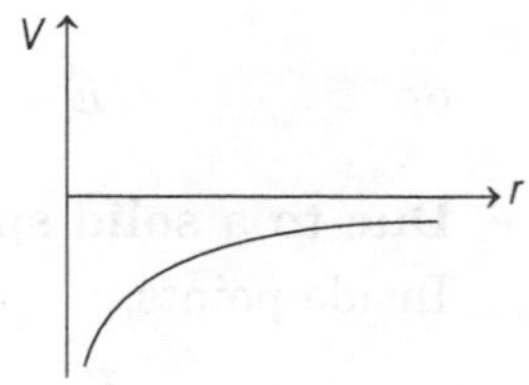

- **Due to a solid sphere**

Inside points $V_i = -\frac{GM}{R^3}(1.5R^2 - 0.5r^2)$

At $r = R$, i.e. on surface $V = -\frac{GM}{R}$

At $r = 0$, i.e. at centre $V = -1.5\frac{GM}{R}$

V-r graph is parabolic for inside points and potential at centre is 1.5 times the potential at surface.

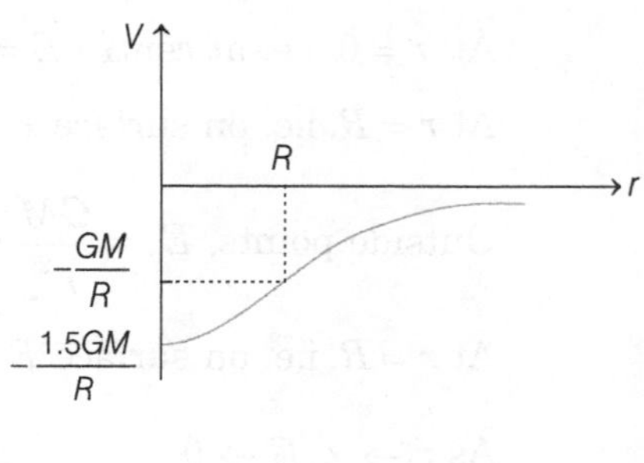

Outside points $V_o = -\dfrac{GM}{r}$

At $r = R$, i.e. on surface $V = -\dfrac{GM}{R}$

As $r \to \infty, V \to 0$

- **Due to a spherical shell**

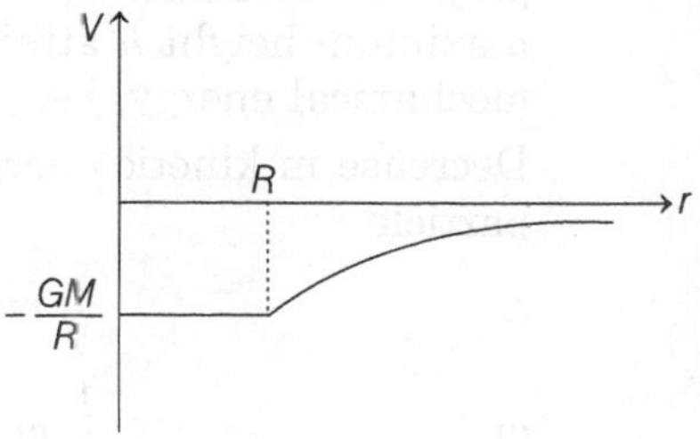

Inside points, $V_i = -\dfrac{GM}{R} = \text{constant}$

Outside points, $V_o = -\dfrac{GM}{r}$

- **On the axis of a ring** $V_x = -\dfrac{GM}{\sqrt{R^2 + x^2}}$

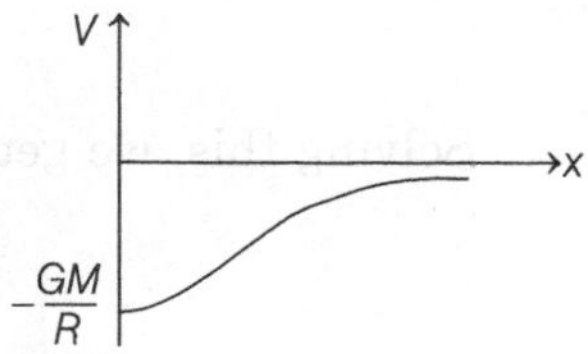

At $x = 0$, i.e. at centre, $V = -\dfrac{GM}{R}$

This is the minimum value.

As $\quad x \to \infty, V \to 0$

Gravitational Potential Energy

- This is negative of work done by gravitational forces in making the system from infinite separation to the present position.
- Gravitational potential energy of two point masses is

$$U = -\frac{Gm_1 m_2}{r}$$

- To find gravitational potential energy of more than two point masses, we have to make pairs of masses but neither of the pair should be repeated. For example, in case of four point masses,

$$U = -G\left[\frac{m_4 m_3}{r_{43}} + \frac{m_4 m_2}{r_{42}} + \frac{m_4 m_1}{r_{41}} + \frac{m_3 m_2}{r_{32}} + \frac{m_3 m_1}{r_{31}} + \frac{m_2 m_1}{r_{21}}\right]$$

For n point masses, total number of pairs will be $\dfrac{n(n-1)}{2}$.

- If a point mass m is placed on the surface of earth, the potential energy here can be given as

$$U_0 = -\frac{GMm}{R}$$

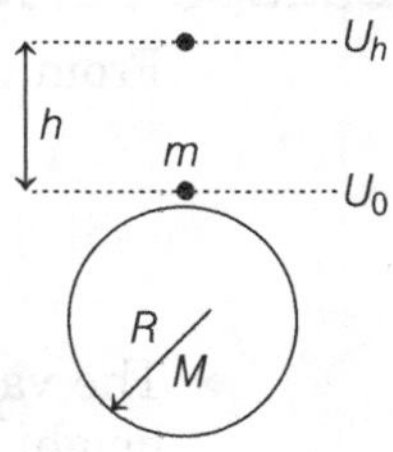

and potential energy at height h is

$$U_h = -\frac{GMm}{(R+h)}$$

The difference in potential energy will be

$$\Delta U = U_h - U_0$$

or $$\Delta U = \frac{mgh}{1 + \frac{h}{R}}$$

If $h << R, \Delta U \approx mgh$

- **Maximum height attained by a particle** Suppose a particle of mass m is projected vertically upwards with a speed v and we want to find the maximum height h attained by the particle. Then, we can use conservation of mechanical energy, i.e.

  Decrease in kinetic energy = increase in gravitational potential energy of particle

  $\therefore$ $$\frac{1}{2} mv^2 = \Delta U$$

  or $$\frac{1}{2} mv^2 = \frac{mgh}{1 + \frac{h}{R}}$$

  Solving this, we get $$h = \frac{v^2}{2g - \frac{v^2}{R}}$$

  From this, we can see that $h \approx \frac{v^2}{2g}$, if v is small.

Relation between Field Strength E and Potential V

***Case* 1** Conversion of V into **E**

(i) If V is a function of only one variable (say r), then

$$E = -\frac{dV}{dr} = -\text{ slope of } V\text{-}r \text{ graph}$$
$$= -dV/dx = -\text{ slope of } V\text{-}x \text{ graph}$$

(ii) If V is a function of more than one variables, then

$$\mathbf{E} = -\left[\frac{\partial V}{\partial x}\hat{\mathbf{i}} + \frac{\partial V}{\partial y}\hat{\mathbf{j}} + \frac{\partial V}{\partial z}\hat{\mathbf{k}}\right]$$

***Case* 2** Conversion of **E** into V

(i) $dV = -\mathbf{E} \cdot d\mathbf{r}$ (More than one variables)

(ii) $dV = -Edr$ or $-Edx$ (One variable)

Escape Velocity

- From the surface of earth,

  $$v_e = \sqrt{2gR} = \sqrt{\frac{2GM}{R}} \qquad \left(\text{as } g = \frac{GM}{R^2}\right)$$
  $$\approx 11.2 \text{ km/s}$$

- The value of escape velocity is 11.2 km/s from the surface of earth. From some height above the surface of earth, this value will be less than 11.2 km/s.

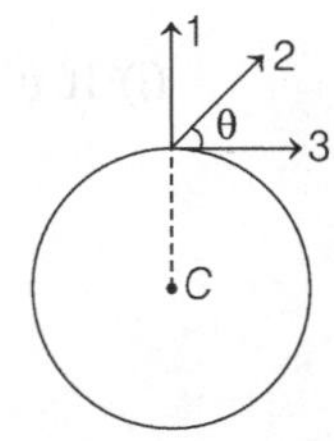

- Escape velocity is independent of the direction in which it is projected. In the figure shown, body is given 11.2 km/s along three different paths. In each case, it will escape to infinity, but following different paths.

  For example, along path-1 it will follow a straight line.
- If velocity of a particle is v_e, then its total mechanical energy is zero. As the particle moves towards infinity, its kinetic energy decreases and potential energy increases, but total mechanical energy remains constant. At any point

$$E = K + U = 0 \Rightarrow K = -U$$

  For example, if $K = 100$ J on the surface of earth, then $U = -100$ J. At some height suppose K becomes 60 J, then U will become -60 J. At infinity $K = 0$, so U is also zero. Hence, speed at infinity will be zero.
- If velocity of the particle is less than v_e, then total mechanical energy is negative and it does not escape to infinity.
- If velocity of the particle is more than v_e, then total mechanical energy is positive. In this case, at infinity some kinetic energy and speed are left in the particle. Although its potential energy becomes zero.

Motion of Satellites

- Orbital speed $v_o = \sqrt{\dfrac{GM}{r}}$
- Time period $T = \dfrac{2\pi}{\sqrt{GM}} r^{3/2}$
- Kinetic energy $K = \dfrac{GMm}{2r}$
- Potential energy $U = -\dfrac{GMm}{r}$
- Total mechanical energy $E = -\dfrac{GMm}{2r}$
- Near the surface of earth, $r \approx R$ and $v_o = \sqrt{\dfrac{GM}{R}} = \sqrt{gR} = \dfrac{v_e}{\sqrt{2}} = 7.9\ \text{kms}^{-1}$.

  This is the maximum speed of any earth's satellite.
- Time period of such a satellite would be

$$T = \frac{2\pi}{\sqrt{GM}} R^{3/2} = 2\pi\sqrt{\frac{R}{g}}$$

$$= 84.6 \text{ min}$$

  This is the minimum time period of any earth's satellite.

Trajectory of a Body Projected from Point *A* in the Direction *AB* with Different Initial Velocities

Let a body be projected from point A with velocity v in the direction AB. For different values of v, the paths are different. Here, the possible cases are:

(i) If $v = 0$, path is a straight line from A to M.

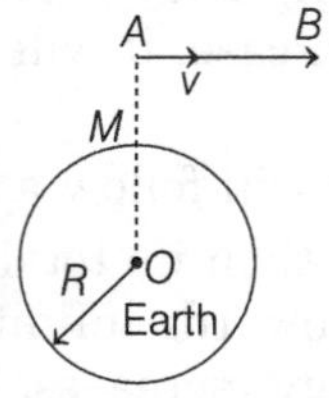

(ii) If $0 < v < v_o$, path is an ellipse with centre O of the earth as a focus.

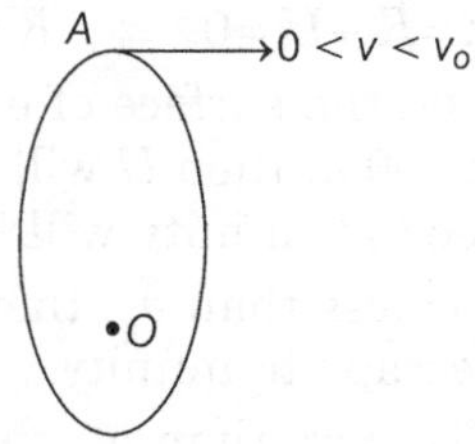

(iii) If $v = v_o$, path is a circle with O as the centre.

(iv) If $v_o < v < v_e$, path is again an ellipse with O as a focus.

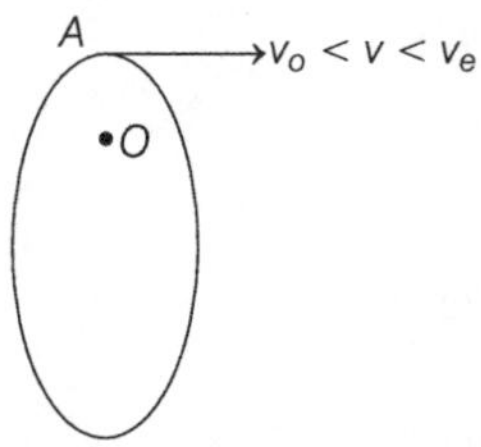

(v) If $v = v_e$, body escapes from the gravitational pull of the earth and path is a parabola.

(vi) If $v > v_e$, body again escapes but now the path is a hyperbola.

Here, v_o = orbital speed $\left(\sqrt{\frac{GM}{r}}\right)$ at A for a circular orbit and v_e = escape velocity from A.

Note *(i) From case (i) to (iv), total mechanical energy is negative. Hence, these are the closed orbits. For case (v), total energy is zero and for case (vi) ,total energy is positive. In these two cases orbits are open.*

(ii) If v is not very large, the elliptical orbit will intersect the earth and the body will fall back to earth.

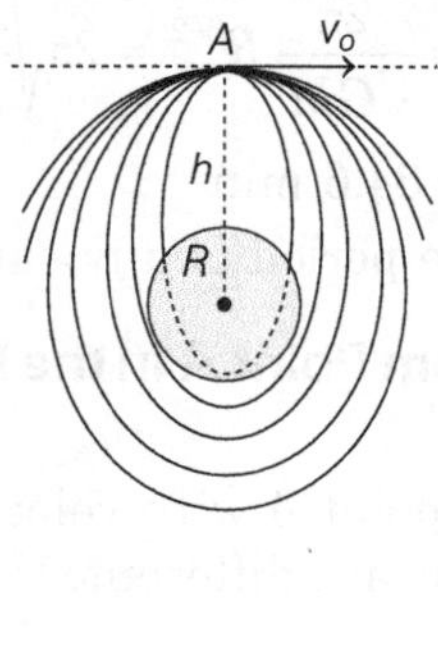

Geostationary or Parking Satellites

A satellite which appears to be at a fixed position at a definite height to an observer on earth is called geostationary or parking satellite. They rotate from west to east.

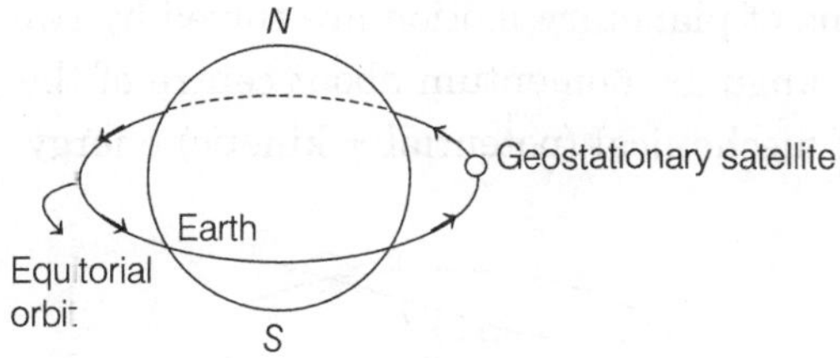

Height from earth's surface = 36000 km

Time period = 24 h

Orbital velocity = 3.1 km/s

$$\text{Angular velocity} = \frac{2\pi}{24} = \frac{\pi}{12}\ \text{rad/h}$$

These satellites are used in communication purpose.

INSAT 2B and INSAT 2C are geostationary satellites of India.

Polar Satellites

These are those satellites which revolve in polar orbits around earth.

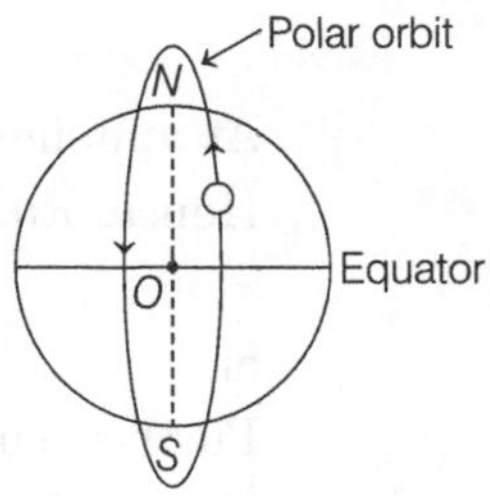

Height from earth's surface ≈ 880 km

Time period ≈ 90 min

Orbital velocity ≈ 8 km/s

$$\text{Angular velocity} \approx \frac{2\pi}{90} = \frac{\pi}{45}\ \text{rad/min}$$

These satellites revolve around the earth in polar orbits.
These satellites are used in forecasting weather, studying the upper region of the atmosphere, in mapping, etc.
PSLV series satellites are polar satellites of India.

Kepler's Laws

- Kepler's three empirical laws describe the motion of planets.

  First law Each planet moves in an elliptical orbit, with the sun at one focus of the ellipse.

  Second law The radius vector drawn from the sun to a planet, sweeps out equal areas in equal time interval, i.e. areal velocity is constant. This law is derived from law of conservation of angular momentum.

$$\frac{dA}{dt} = \frac{L}{2m} = \text{constant.}$$

Here, L is angular momentum and m is mass of planet.

Third law $T^2 \propto r^3$, where r is semi-major axis of elliptical path.

Note *Circle is a special case of an ellipse. Therefore, second and third laws can also be applied for circular path. In third law, r is radius of circular path.*

- Most of the problems of planetary motion are solved by two conservation laws:
  (i) conservation of angular momentum about centre of the sun and
  (ii) conservation of mechanical (potential + kinetic) energy

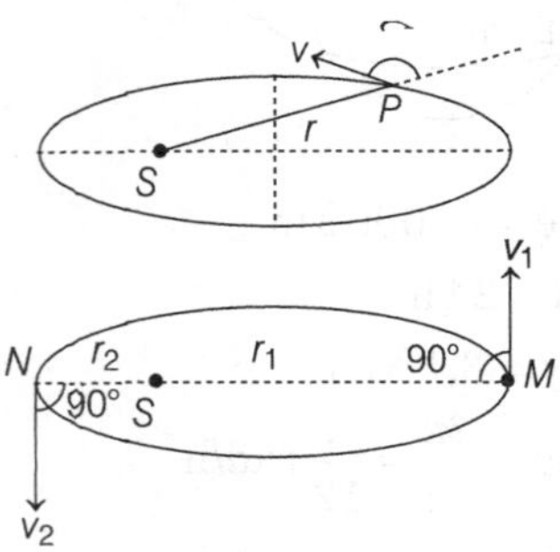

Hence, the following two equations are used in most of the cases,

$$mvr \sin \theta = \text{constant} \qquad \text{...(i)}$$

$$\frac{1}{2} mv^2 - \frac{GMm}{r} = \text{constant} \qquad \text{...(ii)}$$

At aphelion (or M) and perihelion (or N) positions, $\theta = 90°$.

Hence, Eq. (i) can be written as

$$mvr \sin 90° = \text{constant}$$

or
$$mvr = \text{constant} \qquad \text{...(iii)}$$

Further, since mass of the planet (m) also remains constant, so, Eq. (i) can also be written as

$$vr \sin \theta = \text{constant} \qquad \text{...(iv)}$$

or
$$v_1 r_1 = v_2 r_2 \qquad (\because \theta = 90°)$$

$$r_1 > r_2$$

$$\Rightarrow \quad v_1 < v_2$$

- If the law of force obeys the inverse square law, then

$$\left(F \propto \frac{1}{r^2},\ F = \frac{-dU}{dr} \right) \Rightarrow \quad K = \frac{|U|}{2} = |E|$$

The same is true for electron-nucleus system because there also, the electrostatic force $F_e \propto \frac{1}{r^2}$.

CHAPTER 10

Properties of Solids

- **Stress** $= \dfrac{F}{A}$ = restoring force per unit area
- **Strain** $= \dfrac{\Delta x}{x}$ = change in original length per unit
- **Modulus of Elasticity** $E = \dfrac{\text{stress}}{\text{strain}}$

Note *Materials which offer more resistance to external deforming forces have higher value of modulus of elasticity.*

- **Young's Modulus of Elasticity**

$$Y = \frac{F/A}{\Delta l/l} = \frac{Fl}{A\,\Delta l}$$

- **Bulk Modulus of Elasticity**

$$B = \frac{F/A}{\Delta V/V} = -\frac{\Delta p}{\Delta V/V}$$

or $$-\left(\frac{dp}{dV}\right)V$$

- **Shear Modulus of Elasticity or Modulus of Rigidity**

$$\eta = \frac{F/A}{\theta}$$

- Solids have all three modulii of elasticities, Young's modulus, bulk modulus and shear modulus whereas liquids and gases have only bulk modulus.

- Every wire is like a spring whose force constant is equal to $\frac{YA}{l}$, i.e.

$$k = \frac{YA}{l} \quad \text{or} \quad k \propto \frac{1}{l}$$

- Potential energy stored in a stretched wire

$$U = \frac{1}{2} k (\Delta l)^2 = \frac{1}{2}\left(\frac{YA}{l}\right)(\Delta l)^2$$

- Potential energy stored per unit volume (also called energy density) in a stretched wire

$$u = \frac{1}{2} \times \text{stress} \times \text{strain}$$

- **Change in length of a wire** $\Delta l = \frac{F l}{AY}$

  Here, F is tension in the wire. If wire is having negligible mass, tension is uniform throughout the wire and change in length is obtained directly, otherwise by integration.

- In case of solids and liquids, bulk modulus is almost constant. In case of a gas, it is process dependent.

  In isothermal process, $B = B_T = p$

  In adiabatic process, $B = B_S = \gamma p$

- Compressibility $= \frac{1}{B}$

- When pressure is applied on a substance, its volume decreases while mass remains constant. Hence, its density will increase,

$$\Delta \rho = \frac{\rho \Delta p}{B}$$

- In the figure shown,

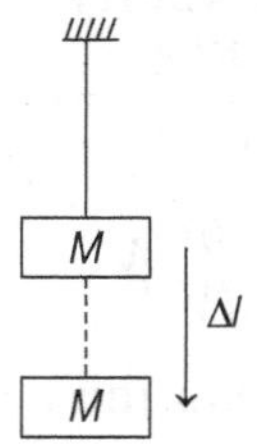

Work done by gravity is $W = (Mg)\Delta l$ ...(i)

But potential energy stored in the stretched wire,

$$U = \frac{1}{2}(Mg)(\Delta l) \quad \text{...(ii)}$$

From Eqs. (i) and (ii), we can see that half of the work done by gravity is stored as potential energy in stretched wire and rest half (or 50%) is dissipated in the form of heat, sound, etc. during stretching.

Stress-strain curve

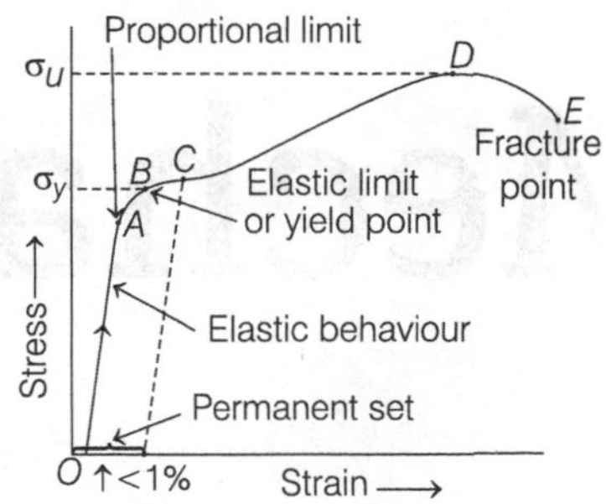

Analysis of the Curve

- In figure, we can see that in the region between O and A, the curve is linear. Hence, Hooke's law (stress ∝ strain) obeys in this region.
- In the region from A to B, the stress and strain are not proportional. However, if we remove the load, the body returns to its original dimension.
  The point B in the curve is the yield point or the elastic limit and the corresponding stress is the yield strength (σ_y) of the material.
- Once the load is increased further, the strain increases rapidly even for a small change in the stress.
  This is shown in the region from B to D in the curve.
- If the load is removed at a point C (say) between B and D, the body does not regain its original dimension. Hence, even when the stress is zero, the strain is not zero and the deformation is called plastic deformation.
- Further the point D is the ultimate tensile strength (σ_u) of the material. Hence, if any additional strain is produced beyond this point, a fracture can occur (point E). If
  (i) the ultimate strength and fracture points are close to each other (points D and E), then the material is brittle.
  (ii) the ultimate strength and fracture points are far apart (points D and E), then the material is ductile.

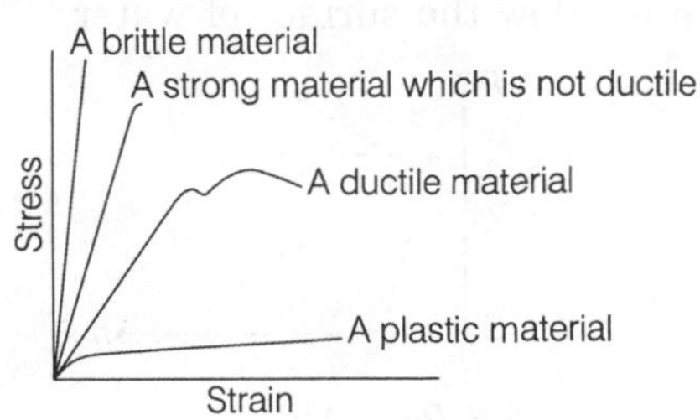

CHAPTER 11

Fluid Mechanics

Pressure and Upthrust

- Upthrust $F = V_i \rho_l g_e$
- When a solid whose density is less than the density of liquid floats in it then, some fraction of solid remains immersed in the liquid. In this case,
  (i) Weight = upthrust
  (ii) Fraction of immersed volume, $f = \frac{\rho_s}{\rho_l}$
- When a solid whose density is more than the density of liquid is completely immersed in it, then upthrust acts on its 100% volume and apparent weight is less than its actual weight.
$$w_{app} = w - F$$
  Here, F = Upthrust on 100% volume of solid.
- Relative density (or specific gravity) of any substance
$$\text{RD} = \frac{\text{density of that substance}}{\text{density of water}} = \frac{\text{weight in air}}{\text{change in weight in water}}$$
- $1\,\text{Pa} = 1\,\text{Nm}^{-2}$, $1\,\text{Bar} = 10^5\,\text{Pa}$, $1\,\text{atm} = 1.013 \times 10^5\,\text{Pa}$
  Gauge pressure = absolute pressure − atmospheric pressure
- Pressure at depth h below the surface of water,

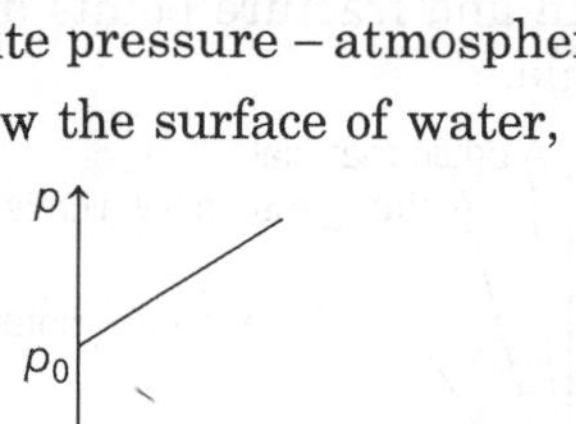

$$p = p_0 + \rho g h$$

Change in pressure per unit depth,
$$\frac{dp}{dh} = \rho g$$

- In the figure given below, liquid pressure will be same at all points at the same level (their provided speeds are same).

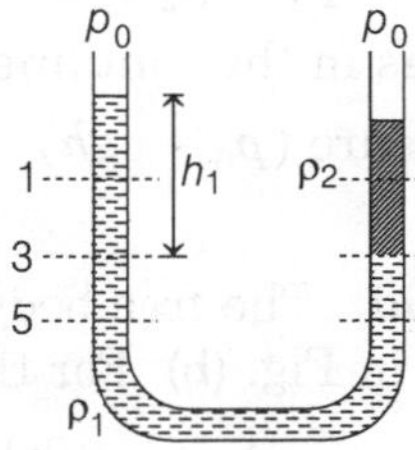

For example, in the figure

$$p_1 \neq p_2,\ p_3 = p_4 \text{ and } p_5 = p_6$$

Further, $p_3 = p_4$

$\therefore$ $p_0 + \rho_1 g h_1 = p_0 + \rho_2 g h_2$

or $\rho_1 h_1 = \rho_2 h_2$ or $h \propto \dfrac{1}{\rho}$

- **Barometer** It is a device used to measure atmospheric pressure.

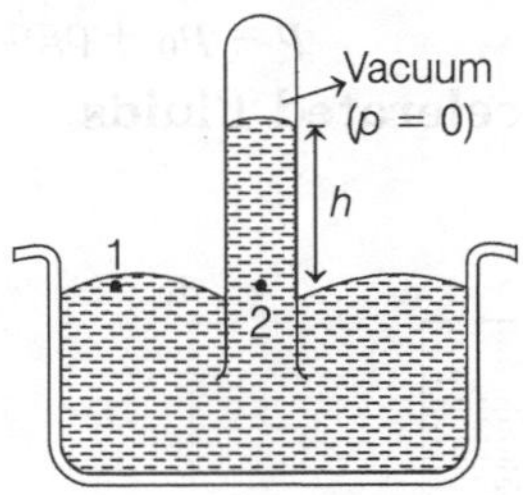

$\because$ $p_1 = p_2$

Here, p_1 = atmospheric pressure (p_0)

and $p_2 = 0 + \rho g h = \rho g h$

Here, ρ = density of mercury

$\therefore$ $p_0 = \rho g h$

Thus, the mercury barometer reads the atmospheric pressure (p_0) directly from the height of the mercury column.

- **Manometer** It is a device used to measure the pressure of a gas inside a container.

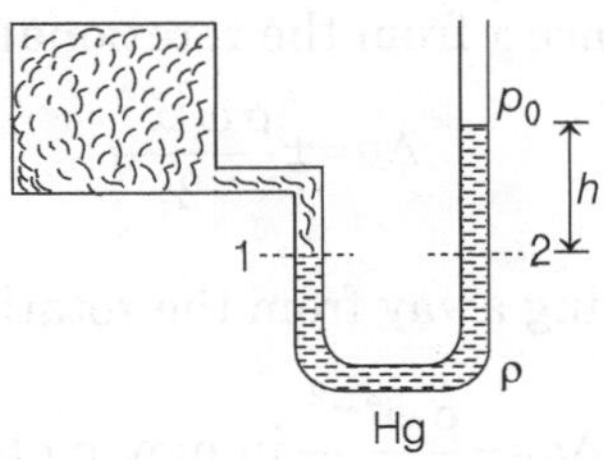

The U-shaped tube often contains mercury

$$p_1 = p_2$$

Here, p_1 = pressure of the gas in the container (p)

and p_2 = atmospheric pressure ($p_0 + \rho gh$)

$\therefore$ $p_{gas} = p_1 = p_0 + h\rho g$

- **Free Body Diagram of a Liquid** The free body diagram of the liquid (showing the vertical forces only) is shown in Fig. (b). For the equilibrium of liquid,

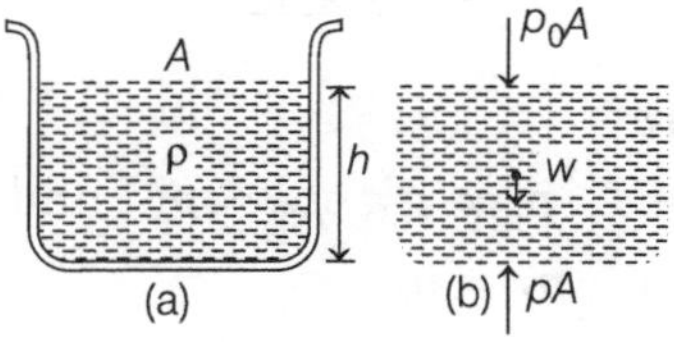

Net downward force = net upward force

$\therefore$ $$p_0 A + W = pA$$

Here, $$W = \rho ghA$$

$\therefore$ $$p = p_0 + \rho gh$$

- **Change in Pressure in Accelerated Fluids**

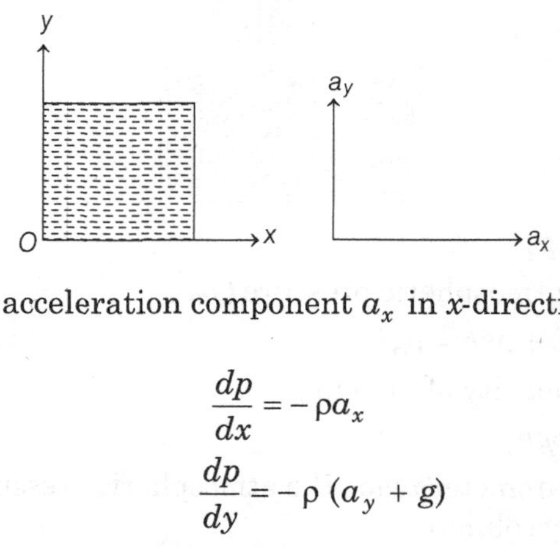

If container has an acceleration component a_x in x-direction and a_y in y-direction.Then,

$$\frac{dp}{dx} = -\rho a_x$$

and $$\frac{dp}{dy} = -\rho (a_y + g)$$

- **Pressure Difference in Rotating Fluids**

In a rotating fluid (also accelerating) pressure increases in moving away from the rotational axis. At a distance x from the rotational axis, the pressure difference is

$$\Delta p = \pm \frac{\rho \omega^2 x^2}{2}$$

Take, $\Delta p = +\frac{\rho \omega^2 x^2}{2}$ in moving away from the rotational axis, as pressure increases in this direction and take $\Delta p = -\frac{\rho \omega^2 x^2}{2}$ in moving towards the rotational axis.

- **Variation of Pressure with Depth** Consider a cylinder kept inside a liquid as shown in figure.

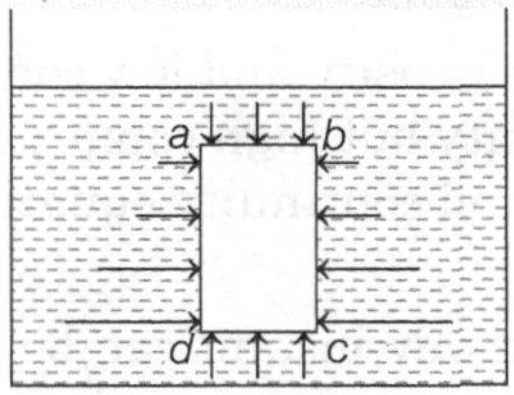

Pressure increases linearly with depth as

$$p = p_0 + \rho g h$$

Therefore, if h is same, then pressure is also same or we can say that on a horizontal surface pressure will be same.

For example, pressure at all points on horizontal surface (ab) will be same. Similarly, pressure at all points on horizontal surface (dc) is also same. But

$$p_{dc} > p_{ab} \qquad (\text{As } h_{dc} > h_{ab})$$

But pressure on vertical faces ad and bc increases linearly with depth as shown in figure.

- **Pressure Force and its Torque**

$$p = \frac{F}{A} \Rightarrow F = pA$$

Let us call this force as the pressure force. Now, this force is calculated on a surface. If surface is horizontal then pressure is uniform at all points.

So, $F = pA$ can be applied directly. For calculation of torque, point of application of force is required. In the above case, point of application of force may be assumed at geometrical centre of the surface.

If the surface is vertical or inclined, pressure is non-uniform (it increases with depth) so pressure force and its torque can be obtained by integration. After finding force and torque by integration, we can also find point of application of this force by the relation,

$$r_\perp = \frac{\tau}{F} \qquad (\text{As } \tau = F \times r_1)$$

Note *For small heights, pressure in a gas (including the atmosphere) does not change much. So, atmospheric pressure P_0 is assumed almost constant unless height difference is large. So, pressure force due to P_0 will be P_0A and which can be found directly. The reason of constant pressure is,*

$$\Delta p = \rho g \Delta h$$

Since, ρ of a gas is negligible. Therefore, ΔP tends to zero or pressure is almost constant.

Ideal Fluid

- An ideal fluid is incompressible and non-viscous. An incompressible fluid cannot be pressed at all by applying pressure on it. Its bulk modulus of elasticity is infinite and compressibility is zero.

Since, its volume cannot be changed, so its density remains constant. A non-viscous fluid offers no internal friction. An object moving through this fluid does not experience a retarding force.

Normally, liquids have high viscosity and low compressibility (compared to gases). Gases have low viscosity but high compressibility. Bernoulli's equation is applicable for an ideal fluid and continuity equation is applicable for an incompressible fluid.

- **Streamlines** A streamline at any instant can be defined as an imaginary curve or line, so that tangent to the curve at any point represents the direction of the instantaneous velocity at that point.

In a laminar flow, the streamlines will be fixed. Every particle passing through P has velocity $\mathbf{v}_1$ and at Q it is $\mathbf{v}_2$.

- **Volume Flow Rate** $\quad Q = Av \quad$ or $\quad \frac{dV}{dt} = Av$

Continuity Equation

$$Q_1 = Q_2 \quad \text{or} \quad \frac{dV_1}{dt} = \frac{dV_2}{dt}$$

or
$$A_1 v_1 = A_2 v_2 \quad \text{or} \quad Av = \text{constant}$$

or
$$v \propto \frac{1}{A}$$

Bernoulli's Equation

$$p + \rho gh + \frac{1}{2}\rho v^2 = \text{constant}$$

or
$$p_1 + \rho gh_1 + \frac{1}{2}\rho v_1^2 = p_2 + \rho gh_2 + \frac{1}{2}\rho v_2^2$$

In Bernoulli's equation, there are three terms; p, $\frac{1}{2}\rho v^2$ and ρgh. Under following three cases, this equation reduces to a two term Bernoulli equation :

***Case* 1** If all points are open to atmosphere, then pressure at every point may be assumed to be constant ($= p_0$) and the Bernoulli equation can be written as,

$$\frac{1}{2}\rho v^2 + \rho gh = \text{constant}$$

At greater heights h, speed v will be less as ρ and g are constants.

***Case* 2** If the liquid is passing through a pipe of uniform cross-section, then from continuity equation (Av = constant), speed v is same at all points. Therefore, the Bernoulli equation becomes

$$p + \rho gh = \text{constant}$$

or
$$p_1 + \rho gh_1 = p_2 + \rho gh_2$$

or $$p_1 - p_2 = \rho g(h_2 - h_1)$$

This is the pressure relation we have already discussed for a fluid at rest. Thus, pressure decreases with height of liquid and increases with depth of liquid.

***Case* 3** If a liquid is flowing in a horizontal pipe, then height h of the liquid at every point may be assumed to be constant. So, the two term Bernoulli equation becomes,

$$p + \frac{1}{2}\rho v^2 = \text{constant}$$

From this equation, we may conclude that pressure decreases at a point where speed increases.

- (i) $v = \sqrt{2gh} = \sqrt{2gh_{top}}$

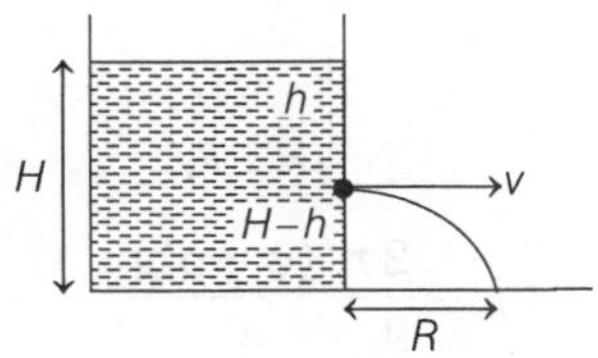

Here, h_{top} = distance of hole from top surface

(ii) $t = \sqrt{\frac{2(H-h)}{g}} = \sqrt{\frac{2h_{bottom}}{g}}$

Here, h_{bottom} = distance of hole from bottom

(iii) $R = vt = 2\sqrt{h(H-h)} = 2\sqrt{h_{top} \times h_{bottom}}$

(iv) $R_{max} = H$ at $h = \frac{H}{2}$

(v) Time taken to empty a tank if hole is made at bottom.

$$t = \frac{A}{a}\sqrt{\frac{2H}{g}}$$

Viscosity

- $F = -\eta A \frac{dv}{dy}$ $\left(\because \frac{dv}{dy} = \text{velocity gradient}\right)$
- For spherical ball, $F = 6\pi\eta r v$
- $v_T = \frac{2}{9}\frac{r^2(\rho - \sigma)g}{\eta}$ or $v_T \propto r^2$

  Here, ρ = density of ball and σ = density of viscous medium in which ball is moving.
- Terminal velocity, $v_T \propto r^2$
- If the fluid is air, then its density σ is negligible compared to density of sphere. So, in that case upthrust will be zero and terminal velocity will be

$$v_T = \frac{2}{9}\frac{r^2 \rho g}{\eta}$$

In this case, weight is equal to the viscous force when terminal velocity is attained.

- If density of fluid is greater than density of sphere ($\sigma>\rho$), then terminal velocity comes out to be negative. So, in this case terminal velocity is upwards.

  In the beginning, upthrust is greater than the weight. Hence, viscous force in this case will be downwards.

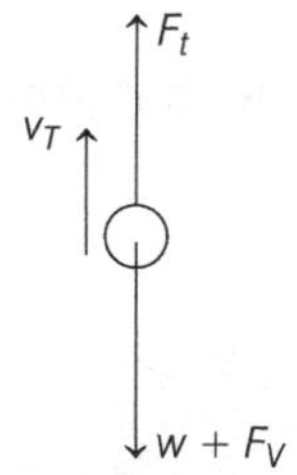

$$v_T = \frac{2}{9}\frac{r^2(\sigma-\rho)g}{\eta}$$

When terminal velocity is attained, then

$$F_t = w + F_v$$

This is the reason why, air bubbles rise up in water.

Surface Tension

- $T = \frac{F}{l} = \frac{\Delta w}{\Delta A}$ or $\Delta w = T \times \Delta A$
- $\Delta p = \frac{2T}{R}$ for single surface and $\frac{4T}{R}$ for double surface
- Capillary rise or fall,

$$h = \frac{2T}{R\rho g} = \frac{2T\cos\theta}{r\rho g} \qquad \left(\text{As } R = \frac{r}{\cos\theta}\right)$$

- **Radius of Curvature of a Curve**

  To describe the shape of a curved surface or interface, it is necessary to know the radii of curvature to a curve at some point. Consider the curve AB as shown in figure

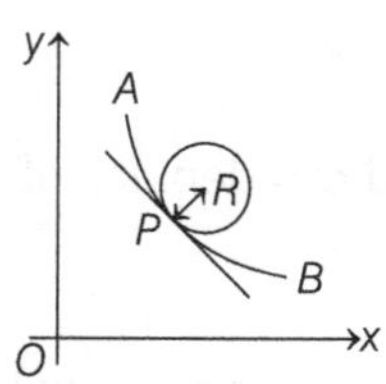

Let P be a point on this curve. The radius of curvature R of AB at P is defined as the radius of the circle which is tangent to the curve at point P.

- **Principal Radii of Curvature of a Surface**
  An infinite set of pairs of radii is possible at any point on a surface. The maximum and minimum radii are called principal radii (denoted by R_1 and R_2).
- **Young-Laplace Equation**
  There exists a difference in pressure across a curved surface which is a consequence of surface tension. The pressure is greater on the concave side. The Laplace equation relates the pressure difference to the shape of the Young surface.
  This difference in pressure is given by

$$\Delta p = T\left(\frac{1}{R_1} + \frac{1}{R_2}\right)$$

For a spherical surface

$$R_1 = R_2 = R \text{ (say), therefore } \Delta p = \frac{2T}{R}$$

For a cylindrical surface

$$R_1 = R \text{ and } R_2 = \infty, \text{ therefore } \Delta p = \frac{T}{R}$$

For a planer surface

$$R_1 = R_2 = \infty \text{ , therefore } \Delta p = 0$$

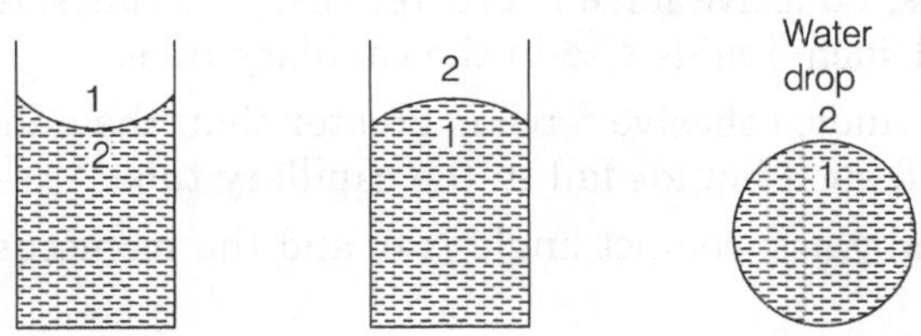

$p_1 - p_2 = \dfrac{2T}{R}$ $\qquad$ $p_1 - p_2 = \dfrac{2T}{R}$ $\qquad$ $p_1 - p_2 = \dfrac{2T}{R}$

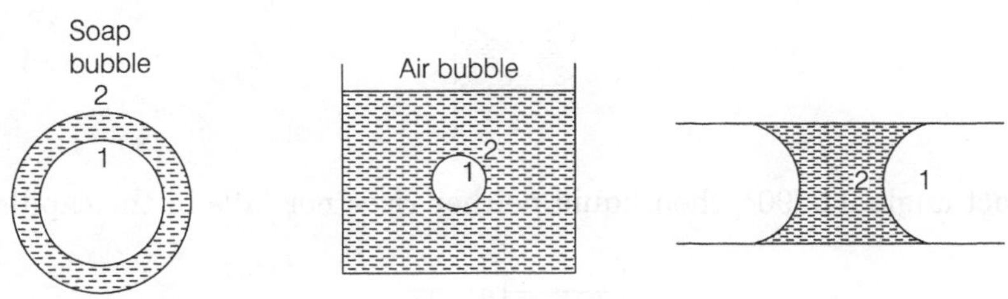

$p_1 - p_2 = \dfrac{4T}{R}$ $\qquad$ $p_1 - p_2 = \dfrac{2T}{R}$ $\qquad$ $p_1 - p_2 = \dfrac{T}{R}$

- In general, surface tension decreases with increase in temperature, because cohesive forces decrease with an increase of molecular thermal activity. Liquid molecules having large intermolecular force will have large surface tension.

- Surface tension of water will increase when highly soluble (like NaCl or sugar) impurities are added otherwise surface tension decreases.
- At critical temperature, surface tension becomes zero.
- **Why Hot Soup is Tastier than Cold Soup?**

  If surface tension is more, liquid surface will have lesser surface area. With increase in temperature, surface tension decreases, so the hot soup spreads over a larger area of tongue and the receptors get more taste than cold soup.

Contact angle (θ)

- The angle between the tangent at the liquid surface at the point of contact (of solid, liquid and gas) and the solid surface inside the liquid is called contact angle θ.

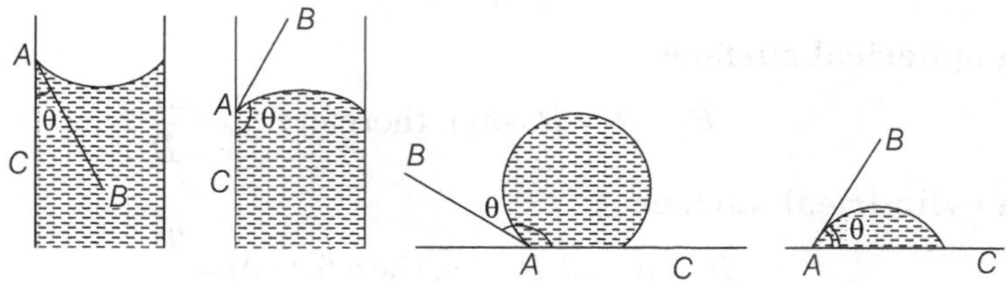

In the four figures shown above, θ is the angle between the lines AB and AC. Here, AB line is away from the solid surface and line AC is towards the liquid.

- For wetting liquids, adhesive force is greater than the cohesive force. Contact angle is acute and such liquids rise in the capillary tube.
- For non-wetting liquids, cohesive force is greater than the adhesive force, contact angle is obtuse and such liquids fall in the capillary tube.
- For pure water and glass, contact angle is 0° and the meniscus is as shown in figure below.

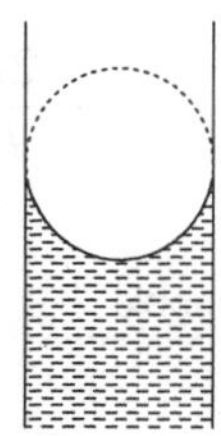

- If contact angle θ is 90°, then liquid neither rises nor falls in the capillary tube.

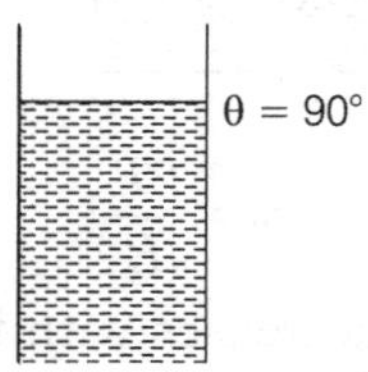

- Shapes of the liquid surface at different contact angles are as shown below.

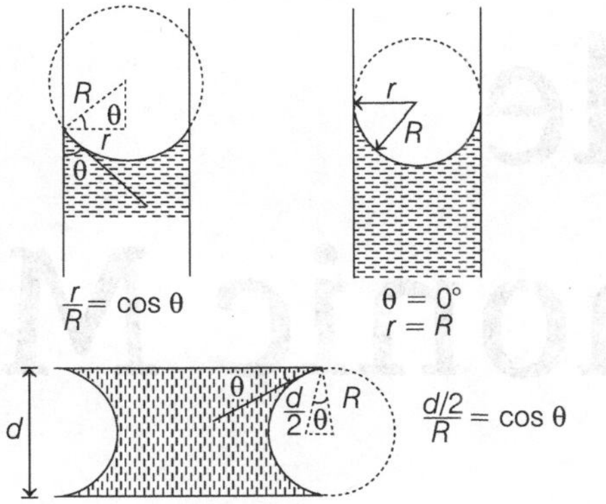

Physical Significance of Reynolds Number

- Consider a narrow tube having a cross-sectional area A. Suppose a fluid flows through it with a velocity v for a time interval Δt.

  Length of the fluid = velocity × time = $v\Delta t$

  Volume of the fluid flowing through the tube in time $\Delta t = Av\Delta t$

  Mass of the fluid,

  $$\Delta m = \text{volume} \times \text{density} = Av\Delta t \times \rho$$

  Inertial force acting per unit area of the fluid

  $$= \frac{F}{A} = \frac{\text{rate of change of momentum}}{A}$$

  $$= \frac{\Delta m \times v}{\Delta t \times A} = \frac{Av\Delta t\rho \times v}{\Delta t \times A} = \rho v^2$$

  Viscous force per unit area of the fluid = η × velocity gradient = $\eta \dfrac{v}{D}$

  $$\frac{\text{Inertial force per unit area}}{\text{Viscous force per unit area}} = \frac{\rho v^2}{\eta v / D}$$

  $$= \frac{\rho v D}{\eta} = \text{Reynold's number } (K)$$

  Thus, Reynold's number represents the ratio of the inertial force per unit area to the viscous force per unit area.

  When $0 < K < 2000$, the flow of liquid is streamlined.

  When $2000 < K < 3000$, the flow of liquid is variable between streamlined and turbulant.

  When $K > 3000$, the flow of liquid is turbulent.

  Reynold's number has no unit and dimension.

CHAPTER 12

Simple Harmonic Motion

Displacement Equations of SHM

- **Force Equation**

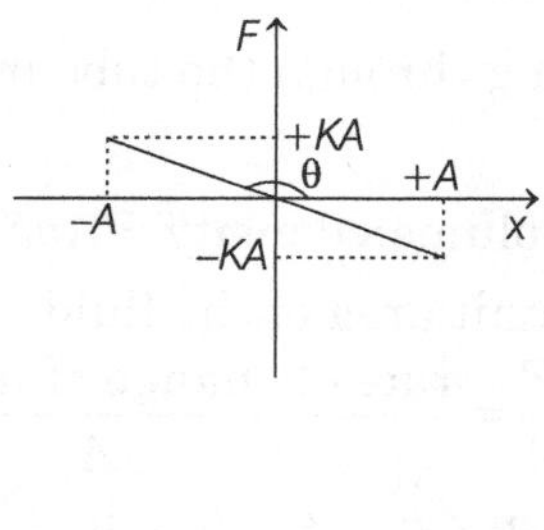

$F = -kx.$

Slope = $\tan\theta = -k$

$|F_{max}| = KA$, at extreme positions or $x = \pm A$

$|F_{min}| = 0$, at mean position, $x = 0$.

- **Acceleration Equation**

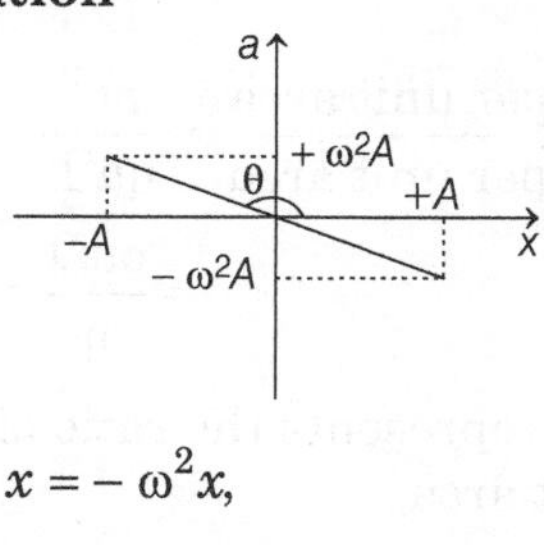

$$a = \frac{F}{m} = -\left(\frac{k}{m}\right)x = -\omega^2 x,$$

$|a|_{min} = 0$ at $x = 0$, at mean position.

$|a|_{max} = \omega^2 A$ at $x = \pm A$, at extreme positions.

$$\omega = \sqrt{\frac{k}{m}} = \text{angular frequency of SHM.}$$

Slope = $\tan\theta = -\omega^2$.

Note *F ∝ – x or a ∝ – x is the sufficient and necessary condition for a periodic motion to be simple harmonic motion.*

- **Velocity Equation**

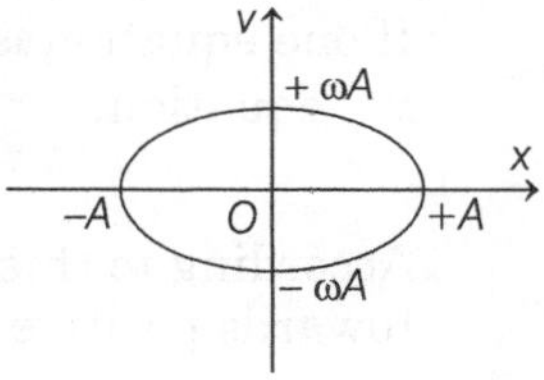

$$v = \pm \omega\sqrt{A^2 - x^2}$$

$$v = 0 \text{ at } x = \pm A$$

or at extreme positions

$v = \pm \omega A$, at $x = 0$ or at mean position.

$|v|_{max} = \omega A$ at mean position and

$|v|_{min} = 0$, at extreme positions.

v versus x graph is an ellipse.

- **Energy Equation**

$$\text{PE} = U_0 + \frac{1}{2}kx^2, \text{KE} = \frac{1}{2}k(A^2 - x^2)$$

and $$\text{TE} = \text{PE} + \text{KE} = U_0 + \frac{1}{2}kA^2$$

Here, U_0 is minimum potential energy at mean position and $\frac{1}{2}kA^2$ or $\frac{1}{2}m\omega^2A^2$ is called energy of oscillation. This much energy is given to the system in the form of KE, PE or both to start oscillations. This much energy keeps on oscillating between potential and kinetic during oscillations.

U_0 or minimum potential energy at mean position may be zero, positive or negative.

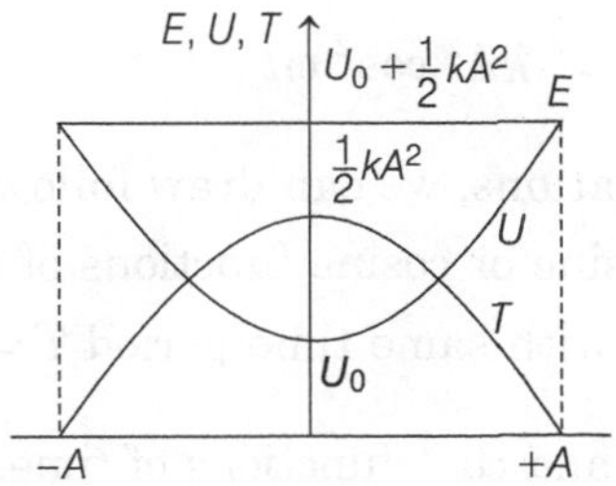

Potential energy *versus x* or kinetic energy *versus x* graph is parabola, while total energy *versus x* graph is a straight line, as it remains constant.

At mean position, $x = 0$.

KE is maximum or $\frac{1}{2}kA^2$ or $\frac{1}{2}m\omega^2A^2$ and potential energy is minimum or U_0.

At extreme positions, $x = \pm A$.

KE is zero and PE is maximum or $U_0 + \frac{1}{2}kA^2$ or $U_0 + \frac{1}{2}m\omega^2A^2$.

Total energy is constant at every point.

This constant energy is $U_0 + \frac{1}{2}kA^2$ or $U_0 + \frac{1}{2}m\omega^2A^2$.

Note *KE can also be represented by T and PE by U.*

Time Equations of SHM

Six time equations in SHM are x - t, v - t, a - t, E - t, U - t and T - t (T =kinetic energy). If one equation is given, other five equations can be made. Let us start with x - t equation,

$$x = A\sin\omega t \quad \text{...(i)}$$

According to this equation, the body starts from mean position and moving towards positive x-direction with initial velocity, $u = +\omega A$.

Now, $$v = \frac{dx}{dt} = \omega A\cos\omega t$$

or $$v = \omega A\cos\omega t \quad \text{...(ii)}$$

$$a = \frac{dv}{dt} = -\omega^2 A\sin\omega t$$

or $$a = -\omega^2 A\sin\omega t \quad \text{...(iii)}$$

$$E = U_0 + \frac{1}{2}kA^2 = U_0 + \frac{1}{2}m\omega^2 A^2 \quad \text{...(iv)}$$

$$U = U_0 + \frac{1}{2}kx^2 = U_0 + \frac{1}{2}kA^2\sin^2\omega t$$

$$= U_0 + \frac{1}{2}m\omega^2 A^2\sin^2\omega t \quad \text{...(v)}$$

$$T = \frac{1}{2}mv^2 = \frac{1}{2}m(\omega A\cos\omega t)^2$$

$$= \frac{1}{2}m\omega^2 A^2\cos^2\omega t$$

$$= \frac{1}{2}kA^2\cos^2\omega t \quad \text{...(vi)}$$

From the above six equations, we can draw following conclusions

(i) x - t, v - t and a - t are sine or cosine functions of same ω. So, x, v and a oscillate sinusoidally with same time period $T = \frac{2\pi}{\omega}$.

(ii) U - t and T - t are $\sin^2$ and $\cos^2$ functions of time. But oscillation frequency of $\sin^2$ or $\cos^2$ functions is double of the oscillation frequency of sine and cosine functions. So, U and T oscillate with double the frequency of oscillations of x, v and a.

Minimum value of U is U_0 and maximum value of U is $U_0 + \frac{1}{2}kA^2$ or $U_0 + \frac{1}{2}m\omega^2 A^2$. Therefore, U oscillates between U_0 and $U_0 + \frac{1}{2}kA^2$. Similarly, minimum value of T is zero and the maximum value is $\frac{1}{2}kA^2$ or $\frac{1}{2}m\omega^2 A^2$.

Hence, T oscillates between 0 and $\frac{1}{2}kA^2$.

(iii) E does not oscillate because it is constant.

(iv) At $t = 0$,

$$x = 0 \qquad \text{(Starts from mean position)}$$

$$v = \omega A \qquad \text{(Maximum velocity)}$$

$$a = 0$$

$$E = U_0 + \frac{1}{2} kA^2$$

$$U = U_0 \qquad \text{(Minimum value)}$$

and
$$T \text{ or KE} = \frac{1}{2} kA^2$$

$$= \frac{1}{2} m \omega^2 A^2 \qquad \text{(Maximum value)}$$

(v) Average value of $\cos^2$ function in one time period is $\frac{1}{2}$. Therefore, from Eq. (vi), we can see that maximum value of kinetic energy is $\frac{1}{2} kA^2$ or $\frac{1}{2} m \omega^2 A^2$.

But average value of kinetic energy in one time period is $\frac{1}{4} kA^2$ or $\frac{1}{4} m \omega^2 A^2$.

(vi) The six graphs are as shown below.

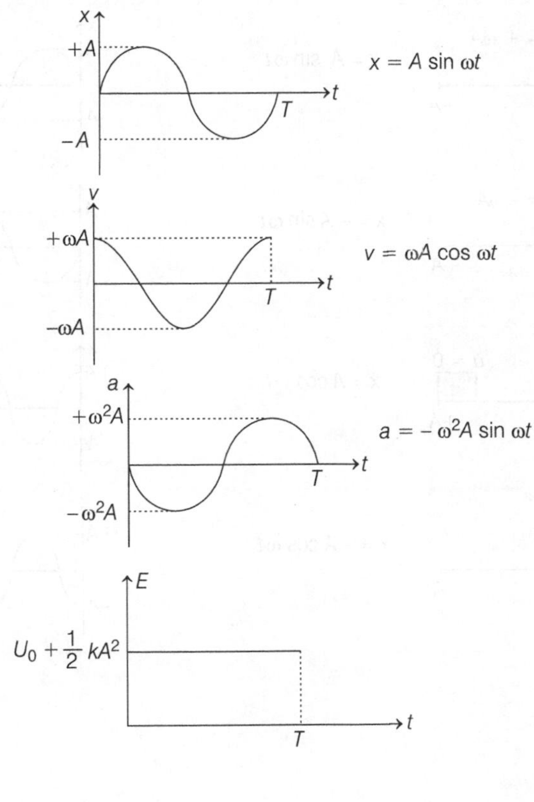

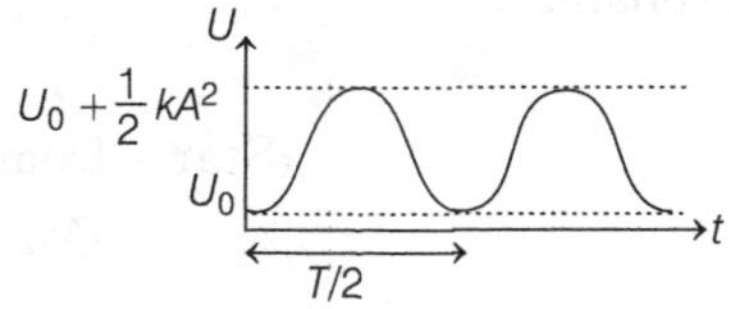

$U = U_0 + \frac{1}{2}kA^2 \sin^2 \omega t$

$KE = \frac{1}{2}kA^2 \cos^2 \omega t$

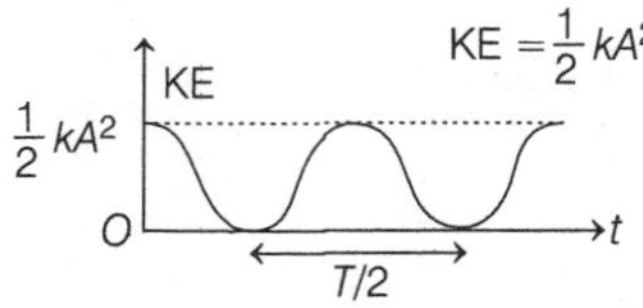

In the first five graphs, we can see that in time $T(=2\pi/\omega)$; x, v and a oscillate once but U and KE oscillate twice.

Note *(i) If the first equation, $x = A \sin \omega t$ is changed, then other five equations (and their graphs also) will change. But the basic nature will remain same. Here, the basic nature is that x, v and a oscillate sinusoidally with same time period and frequency. E does not oscillate. U and KE oscillate with double frequency.*

(ii) Phase diffe1rence between x-t and v-t equations or between v-t and a-t equations is $\frac{\pi}{2}$ or 90°. Phase difference between x-t and a-t equations is π or 180°.

Four Frequently Used *x*-*t* Equations

Initial Conditions at $t = 0$	x-t equation	x-t graph
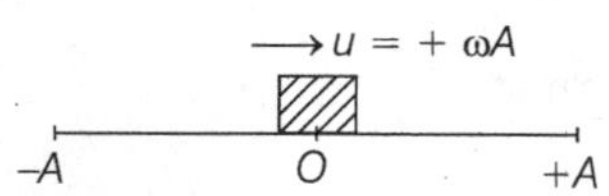	$x = A \sin \omega t$	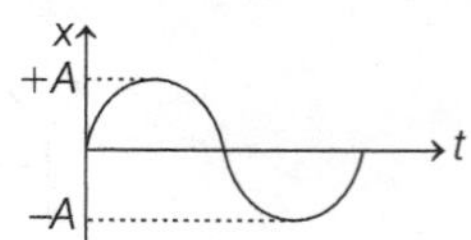
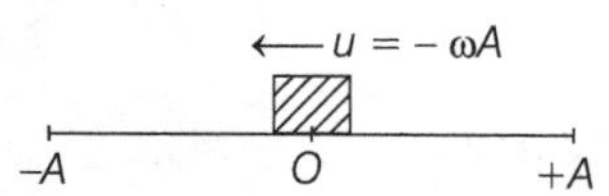	$x = -A \sin \omega t$	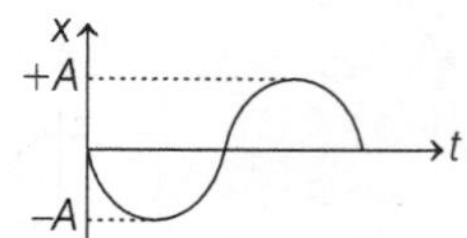
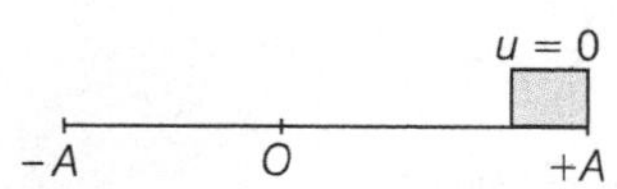	$x = A \cos \omega t$	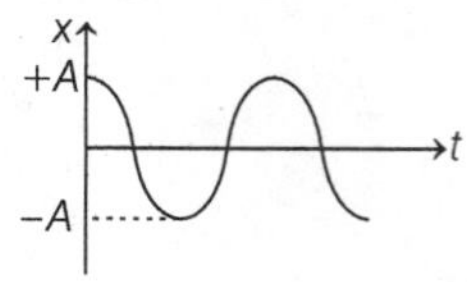
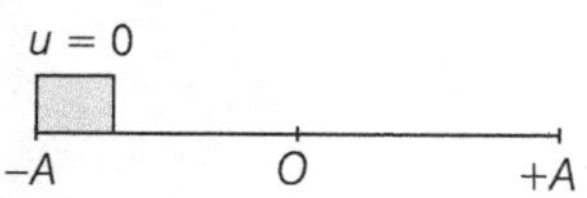	$x = -A \cos \omega t$	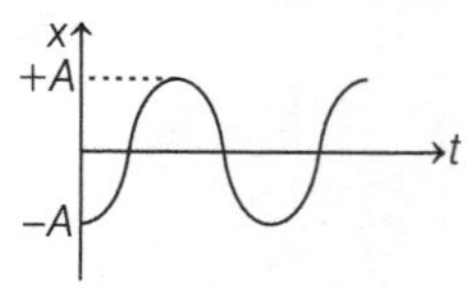

Reference Circle

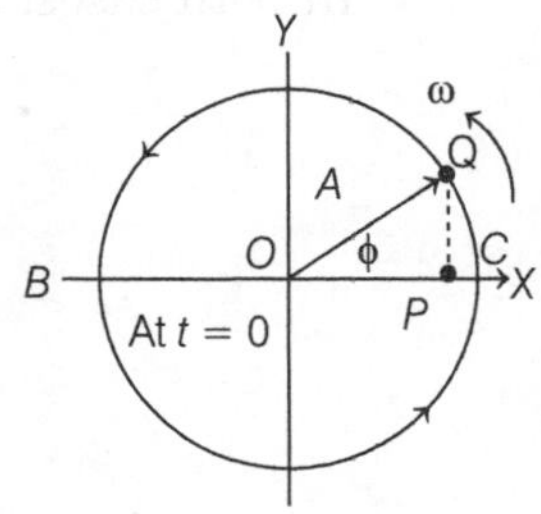

If a particle (say P) rotates in a circle with constant angular speed ω and during the motion of P, we draw a perpendicular from P on any diameter and let this cuts the diameter at Q. Then motion of Q comes out to be simple harmonic.

Angular speed of P becomes the angular freqency of Q. Radius of P becomes the amplitude of Q and centre of P becomes the mean position of Q.

Time Period in SHM

In linear SHM, F-x equation should be like,

$$F = -kx$$

and in angular SHM, τ - θ equation should be like,

$$\tau = -k\theta$$

Then, $$T = 2\pi\sqrt{\frac{m}{k}} \quad \text{(Linear SHM)}$$

and $$T = 2\pi\sqrt{\frac{I}{k}} \quad \text{(Angular SHM)}$$

Spring Block System

- $\omega = \sqrt{\frac{k}{m}}, T = \frac{2\pi}{\omega} = 2\pi\sqrt{\frac{m}{k}}, f = \frac{1}{T} = \frac{1}{2\pi}\sqrt{\frac{k}{m}}$
-

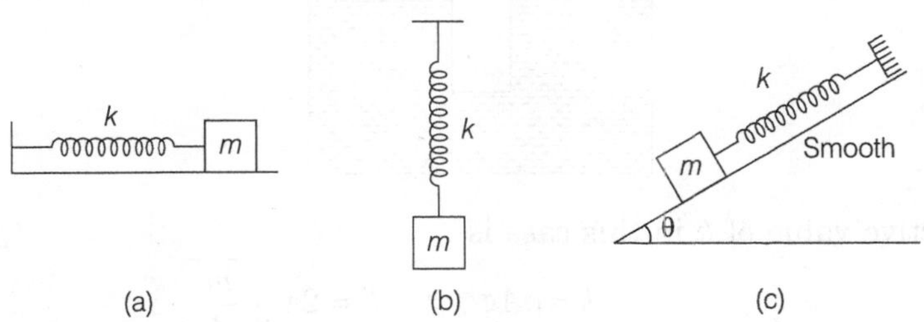

In all three figures shown above, restoring force in displaced position is $F = -kx$. Therefore, time period is,

$$T = 2\pi\sqrt{\frac{m}{k}}$$

The only difference is, their mean positions are different.

In the first figure, mean position is at the natural length of spring.

In the second figure, mean position is obtained after an extension of $x_0 = \frac{mg}{k}$ (as $kx_0 = mg$) and in the third figure mean position is obtained after an extension of $x_0 = \frac{mg\sin\theta}{k}$.

- In both cases, $k_e = k_1 + k_2$

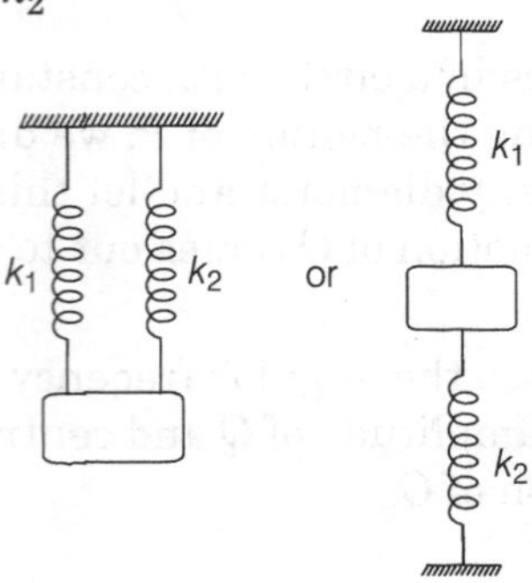

- For the following figure, $\frac{1}{k_e} = \frac{1}{k_1} + \frac{1}{k_2}$

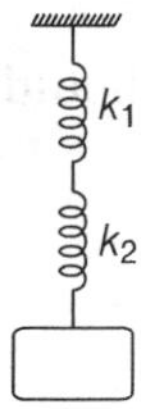

- A plank of mass m and area of cross-section A is floating in a liquid of density ρ. When depressed, it starts oscillating like a spring-block system.

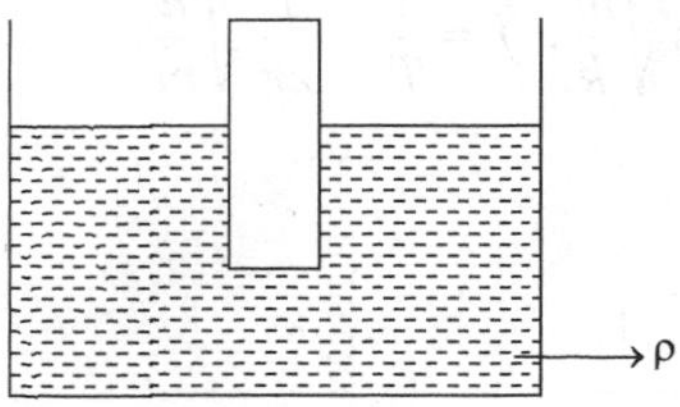

Effective value of k in this case is

$$k = \rho A g \Rightarrow T = 2\pi\sqrt{\frac{m}{\rho A g}}$$

- If mass of spring m_S is also given, then

$$T = 2\pi\sqrt{\frac{m + \frac{m_S}{3}}{k}}$$

- Every wire is also like a spring of force constant given by $k = \frac{YA}{l}$.
- Force constant of a spring is inversely proportional to its length. If length of spring is halved, its force constant will become two times.

Pendulum

- Only small oscillations of a pendulum are simple harmonic in nature.

  Time period of which is given by, $T = 2\pi\sqrt{\frac{l}{g}}$.

Note *Here, l is taken from CM. If liquid is filled in the bob as shown, which is leaking, then l and T will first increase and then decrease.*

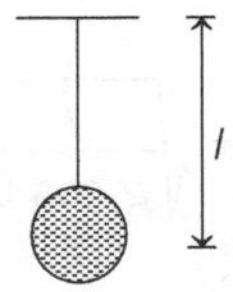

- Second's pendulum is one whose time period is 2 s and length is 1 m.
- For a pendulum having length equivalent to the radius of earth, time period is given as

$$T = 2\pi\sqrt{\frac{1}{g\left(\frac{1}{l}+\frac{1}{R}\right)}}$$

Hence, time period of a pendulum of infinite length is $2\pi\sqrt{\frac{R}{g}}$ or 84.6 min.

Further, $T = 2\pi\sqrt{\frac{l}{g}}$, if $l << R$

or $\frac{1}{l} >> \frac{1}{R}$.

- If point of suspension has an acceleration **a**, then

$$T = 2\pi\sqrt{\frac{l}{|\mathbf{g}_e|}}$$

Here, $\mathbf{g}_e = \mathbf{g} - \mathbf{a} = \mathbf{g} + (-\mathbf{a})$

For example, if point of suspension has an upward acceleration **a**, then (−**a**) is downwards or parallel to **g**.

Hence, $|\mathbf{g}_e| = g + a$

or $T = 2\pi\sqrt{\frac{l}{g+a}}$

- If a constant force **F** (in addition to weight and tension) act on the bob, then

$$T = 2\pi\sqrt{\frac{l}{|\mathbf{g}_e|}}$$

Here, $\mathbf{g}_e = \mathbf{g} + \frac{\mathbf{F}}{m}$

- Above the surface of earth, below the surface of earth, towards equator and on the surface of moon, g decreases, therefore T increases.

- When simple pendulum is in a horizontally accelerated vehicle, then its time period is given by

$$T = 2\pi \sqrt{\frac{l}{\sqrt{(a^2 + g^2)}}}$$

where, a = horizontal acceleration of the vehicle.

- When simple pendulum is in a vehicle sliding down an inclined plane, then its time period is given by

$$T = 2\pi \sqrt{\frac{l}{g \cos \theta}}$$

where, θ = inclination of plane.

Physical Pendulum

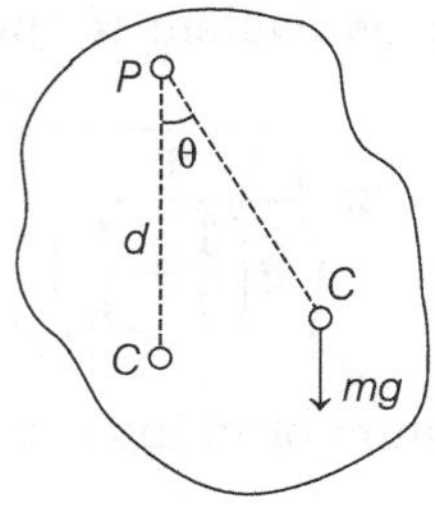

When a rigid body of any shape is capable of oscillating about an axis, it constitutes a physical pendulum and its time period, $T = 2\pi \sqrt{\frac{I}{mgd}}$

The simple pendulum whose time period is same as that of a physical pendulum is termed as an equivalent simple pendulum.

$$T = 2\pi \sqrt{\frac{I}{mgd}} = 2\pi \sqrt{\frac{l}{g}}$$

The length of an equivalent simple pendulum is given by $l = \frac{I}{md}$

Vector Method of Combining Two or More Simple Harmonic Motions

A simple harmonic motion is produced when a force (called restoring force) proportional to the displacement (from the mean position) acts on a particle. If a particle is acted upon by two such forces, the resultant motion of the particle is a combination of two simple harmonic motions.

Suppose the two individual motions are represented by

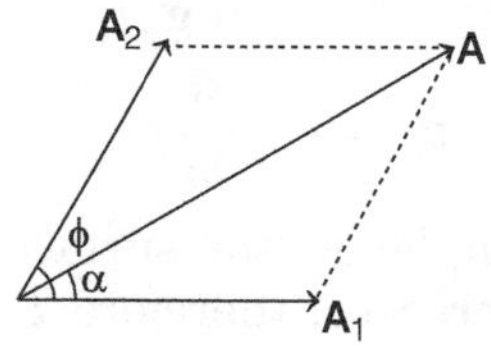

$$x_1 = A_1 \sin \omega t \quad \text{and} \quad x_2 = A_2 \sin (\omega t + \phi)$$

Both the simple harmonic motions have same angular frequency ω.
The resultant displacement of the particle is given by

$$x = x_1 + x_2 = A_1 \sin \omega t + A_2 \sin (\omega t + \phi)$$
$$= A \sin (\omega t + \alpha)$$

Here, $A = \sqrt{A_1^2 + A_2^2 + 2A_1 A_2 \cos \phi}$ and $\tan \alpha = \dfrac{A_2 \sin \phi}{A_1 + A_2 \cos \phi}$

Thus, we can see that this is similar to the vector addition. The same method of vector addition can be applied to the combination of more than two simple harmonic motions.

Two Body Oscillator

If the spring connected with two masses m_1 and m_2 is compressed or elongated by x_0 and then left for oscillations, then both blocks execute SHM with same time period (and therefore, same frequency) but different amplitudes.

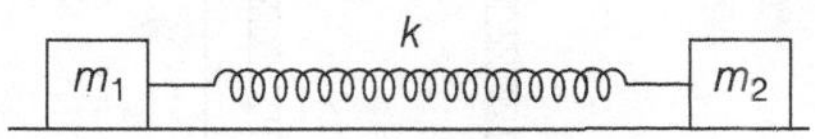

This time period is $$T = 2\pi \sqrt{\frac{\mu}{k}}$$

Here, μ is called their reduced mass given by

$$\frac{1}{\mu} = \frac{1}{m_1} + \frac{1}{m_2} \quad \text{or} \quad \mu = \frac{m_1 m_2}{m_1 + m_2}$$

Their amplitudes of oscillations are in inverse ratio of their masses or

$$A \propto \frac{1}{m} \Rightarrow \frac{A_1}{A_2} = \frac{m_2}{m_1}$$

$\Rightarrow$ $$A_1 = \left(\frac{m_2}{m_1 + m_2}\right) x_0$$

and $$A_2 = \left(\frac{m_1}{m_1 + m_2}\right) x_0$$

Note *If $m_1 >> m_2$, then $\frac{1}{m_1} << \frac{1}{m_2}$*

$\therefore$ $\frac{1}{\mu} \approx \frac{1}{m_2}$ *or* $\mu \approx m_2$ *and* $T = 2\pi\sqrt{\frac{m_2}{k}}$

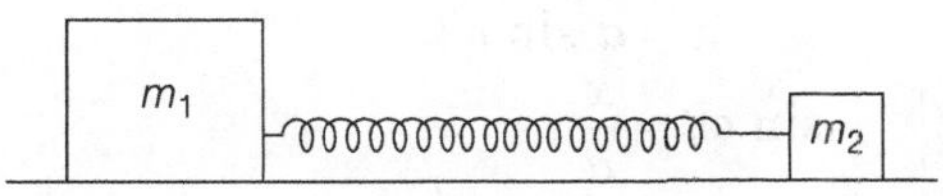

So, in this case, we can imagine that m_1 is almost stationary. Only m_2 will oscillate.

Oscillations of a Fluid Column

- For a fluid column (as shown below), time period is given as

$$T = 2\pi \sqrt{\frac{m}{\rho g A (\sin \theta_1 + \sin \theta_2)}}$$

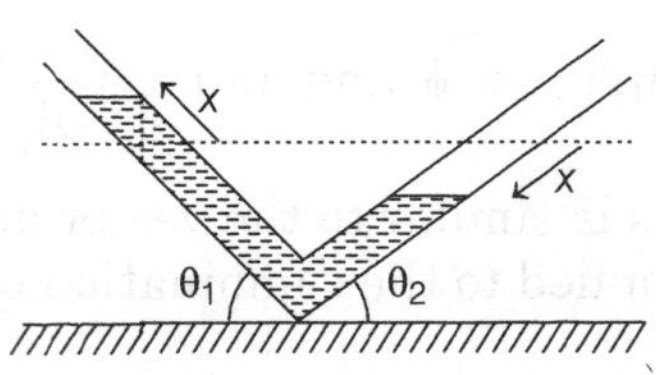

- For a U-tube, if the liquid is filled to a height l, $\theta_1 = 90° = \theta_2$ and $m = 2\,(lA\rho)$

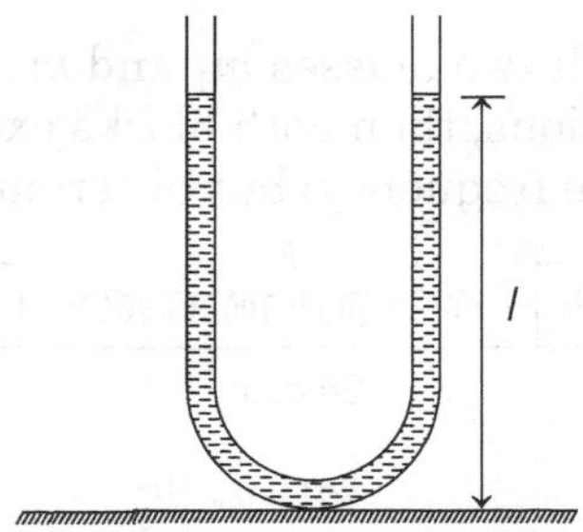

So, $$T = 2\pi \sqrt{\frac{l}{g}}$$

Lissajous Figures

Suppose two forces act on a particle, the first alone would produce a simple harmonic motion in x-direction, given by

$$x = a \sin \omega t$$

and the second would produce a simple harmonic motion in y-direction, given by

$$y = b \sin (\omega t + \phi)$$

The amplitudes a and b may be different and their phases differ by ϕ. The frequencies of the two simple harmonic motions are assumed to be equal.

The resultant motion of the particle is a combination of the two simple harmonic motions.

Depending on the value of ϕ and relation between a and b, the particle follows different paths. These different paths are called Lissajous figures. Given below are few special cases :

***Case* 1 (When $\phi = 0°$)** When the phase difference between two simple harmonic motions is 0°, i.e. $x = a \sin \omega t$

$$\Rightarrow \quad \sin \omega t = \frac{x}{a} \quad \text{...(i)}$$

and $$y = b \sin \omega t$$

$$\Rightarrow \quad \sin \omega t = \frac{y}{b} \quad \text{...(ii)}$$

From Eqs. (i) and (ii), we get

$$\frac{x}{a} = \frac{y}{b} \quad \text{or} \quad y = \left(\frac{b}{a}\right)x$$

which is equation of a straight line with slope $\frac{b}{a}$. Thus, the path of the particle is a straight line. As a special case $y = x$ if $a = b$ or slope is 1.

***Case* 2** $\left(\textbf{When } \phi = \frac{\pi}{2}\right)$ When the phase difference is $\frac{\pi}{2}$, i.e.

$$x = a \sin \omega t \Rightarrow \sin \omega t = \frac{x}{a} \quad \text{...(iii)}$$

and

$$y = b \sin\left(\omega t + \frac{\pi}{2}\right) = b \cos \omega t \Rightarrow \cos \omega t = \frac{y}{b} \quad \text{...(iv)}$$

Squaring and adding, Eqs. (iii) and (iv), we get

$$\frac{x^2}{a^2} + \frac{y^2}{b^2} = 1$$

which is an **ellipse.** Again as a special case, the above equation reduces to,

$$\frac{x^2}{a^2} + \frac{y^2}{a^2} = 1 \quad \text{or} \quad x^2 + y^2 = a^2 \quad (\text{For } a = b)$$

This is an equation of a **circle.**

Damped Oscillations

- The mechanical energy in a real oscillating system decreases during oscillations because of frictional and viscous forces and mechanical energy converts into thermal energy.

  The real oscillator and its motion are then said to be damped. If the damping (or frictional) force is given by $f_r = -bv$, where v is the velocity of the oscillator and b is a damping constant, then the displacement of the oscillator is given by

  $$x(t) = A e^{-\frac{bt}{2m}} \cos(\omega t + \phi)$$

  where, ω is the angular frequency of the damped oscillator and is given by

  $$\omega = \sqrt{\frac{k}{m} - \frac{b^2}{4m^2}}$$

  If the damping constant is small, then $\omega \approx \sqrt{\frac{k}{m}}$, the angular frequency of the undamped oscillator. The mechanical energy E of the damped oscillator is given by

  $$E = \frac{1}{2} kA^2 e^{-bt/m}$$

- If an external force with angular frequency ω_D acts on an oscillating system with natural angular frequency ω, the system oscillates with angular frequency ω_D. The amplitude of oscillations is maximum when $\omega_D = \omega$, such a condition called resonance.

CHAPTER 13

Wave Motion

In any type of wave, oscillations of a physical quantity y are produced at one place and these oscillations (along with energy and momentum) are transferred to other places also.

Classification of Waves

A wave may be classified in following three ways

(i) **Transverse waves** A wave in which the particles of the medium vibrate at right angles to the direction of propagation of wave, is called a transverse wave.

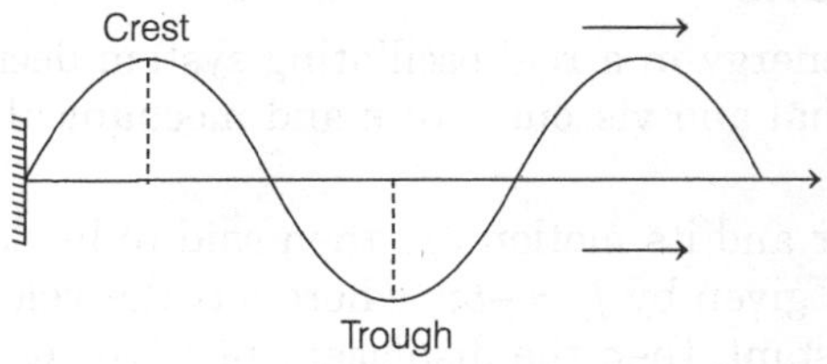

These waves travel in the form of crests and troughs.

It is one dimensional. Transverse waves on the surface of water are two dimensional. Sound wave due to a point source is three dimensional.

(ii) **Longitudinal waves** A wave in which the particles of the medium vibrate in the same direction in which wave is propagating, is called a longitudinal wave.

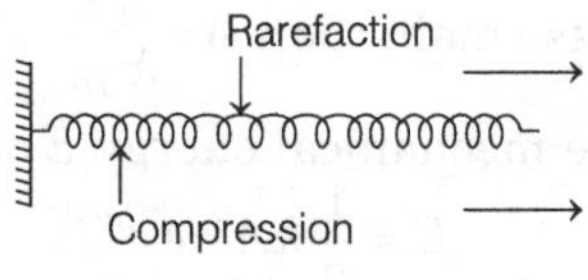

These waves travel in the form of compressions and rarefactions.

(iii) **Mechanical waves** It require medium for their propagation. Sound waves are mechanical in nature. Non-mechanical waves do not require medium for their propagation. Electromagnetic waves are non-mechanical in nature.

Wave Equation

In any wave equation, value of y is a function of position and time. In case of one dimensional wave, position can be represented by one co-ordinate (say x) only. Hence,

$$y = f(x, t)$$

Only those functions of x and t represent a wave equation which satisfy following condition.

$$\frac{\partial^2 y}{\partial x^2} = (\text{constant}) \frac{\partial^2 y}{\partial t^2}$$

Here $\quad$ constant $= \dfrac{1}{v^2}$

where, v is the wave speed.

All functions of x and t of type, $y = f(ax \pm bt)$ satisfy above mentioned condition of wave equation, provided value of y should be finite for any value of x and t. If $y(x, t)$ function is of this type, then following two conclusions can be drawn.

(i) Wave speed, $v = \dfrac{\text{coefficient of } t}{\text{coefficient of } x} = \dfrac{b}{a}$

(ii) Wave travels along positive x-direction, if ax and bt have opposite signs and it travels along negative x-direction, if they have same signs.

Sine Wave

- If oscillations of y are simple harmonic in nature, then wave is called sine wave. General equation of this wave is,

$$y = A\sin(\omega t \pm kx \pm \phi) \quad \text{or} \quad y = A\cos(\omega t \pm kx \pm \phi)$$

In these equations,

- A is amplitude of oscillation,
- ω is angular frequency,

$$T = \frac{2\pi}{\omega}, \ \omega = 2\pi f \ \text{ and } \ f = \frac{1}{T} = \frac{\omega}{2\pi}$$

- k is wave number, $k = \dfrac{2\pi}{\lambda}$ (where, λ = wavelength)
- Wave speed, $v = \dfrac{\omega}{k} = f\lambda$
- ϕ is initial phase angle of the particle at $x = 0$ and
- $(\omega t \pm kx \pm \phi)$ is phase angle at time t of the particle at coordinate x.
- Alternate expressions of a sine wave travelling along positive x-direction are

$$y = A \sin k(x - vt) = A \sin (kx - \omega t)$$
$$= A \sin \frac{2\pi}{\lambda}(x - vt) = A \sin 2\pi\left(\frac{x}{\lambda} - \frac{t}{T}\right)$$

Similarly, the expression,

$$y = A \sin k(x + vt) = A \sin (kx + \omega t)$$

$$= A \sin 2\pi\left(\frac{x}{\lambda} + \frac{t}{T}\right) \text{ etc.}$$

represent a sine wave travelling along negative x-direction.

- **Difference between two equations**

$$y = A\sin(kx - \omega t) \text{ and } y = A\sin(\omega t - kx)$$

Both equations represent a wave travelling in positive x-direction with speed $v = \frac{\omega}{k}$.

The phase difference between them is π. It means, if a particle at $x = 0$ and at time $t = 0$ is at mean position and moving upwards (represented by first wave), then the same particle will be at its mean position but moving downwards (represented by the second wave).

Particle Speed (v_p) and Wave Speed (v) in Case of Sine Wave

- In $y = f(x, t)$ equation, x and t are two variables.

So, $$v_p = \frac{\partial y}{\partial t}$$

- In sine wave, particles are in SHM. Therefore, all equations of SHM can be applied for particles also.
- **Relation between v_p and v**

$$v_p = -v \cdot \frac{\partial y}{\partial x}$$

Phase Difference ($\Delta\phi$)

***Case* I** $\Delta\phi = \omega(t_1 \sim t_2)$

or $$\Delta\phi = \frac{2\pi}{T} \cdot \Delta t$$

= phase difference of one particle at a time interval of Δt.

***Case* II** $\Delta\phi = k(x_1 \sim x_2) = \frac{2\pi}{\lambda} \cdot \Delta x$

= phase difference at one time between two particles at a path difference of Δx.

Energy Density (u), Power (P) and Intensity (I) in Sine Wave

- Energy density, $u = \frac{1}{2}\rho\omega^2 A^2$ = energy of oscillation per unit volume.
- Power, $P = \frac{1}{2}\rho\omega^2 A^2 S v$ = energy transferred per unit time.
- Intensity, $I = \frac{1}{2}\rho\omega^2 A^2 v$ = energy transferred per unit time per unit area.
- For a string segment, the potential energy depends on the slope of the string and is maximum when the slope is maximum, which is at the equilibrium position of the segment, the same position for which the kinetic energy is maximum.

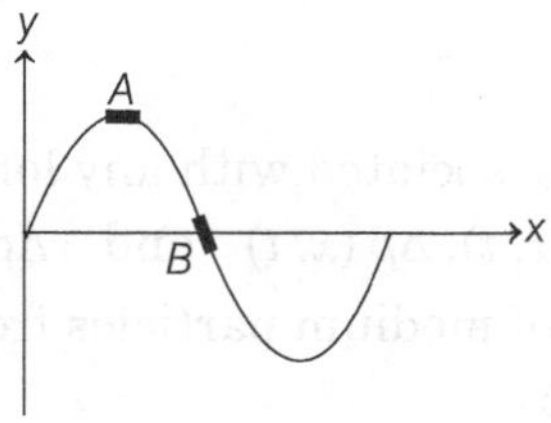

At A : Kinetic energy and potential energy both are zero.
At B : Kinetic energy and potential energy both are maximum.

Note ***Intensity due to a point source*** *If a point source emits wave uniformly in all directions, the energy at a distance r from the source is distributed uniformly on a spherical surface of radius r and area $S = 4\pi r^2$.*

$$I = \frac{P}{S} = \frac{P}{4\pi r^2} \quad or \quad I \propto \frac{1}{r^2}$$

Sound Waves

- Sound waves are the mechanical waves that occur in nature.
- Sound waves are of three types
  (i) **Infrasonic Waves** The sound waves of frequency lying between 0 to 20 Hz are called infrasonic waves.
  (ii) **Audiable Waves** The sound waves of frequency lying between 20 Hz to 20000 Hz are called audiable waves.
  (iii) **Ultrasonic Waves** The sound waves of frequency greater than 20000 Hz are called ultrasonic waves.
  Sound waves are mechanical longitudinal waves and require medium for their propagation. Sound waves can travel through any material medium (i.e. solids, liquid and gases) with speed that depends on the properties of the medium.
- Sound waves cannot propagate through vacuum.
- If v_s, v_l and v_g are speed of sound waves in solid, liquid and gases, then

$$v_s > v_l > v_g$$

- Sound waves (longitudinal waves) can reflect, refract, interfere and diffract but cannot be polarised as only transverse waves can be polarised.

Characteristics of Musical Sound

Musical sound has three characteristics

(i) **Intensity or Loudness** Intensity of sound is energy transmitted per second per unit area by sound waves. Its SI unit is watt/metre^2. Loudness which is related to intensity of sound is measured in decibel (dB).

(ii) **Pitch or Frequency** Pitch of sound directly depends upon frequency.
A shrill and sharp sound has higher pitch and a grave and dull sound has lower pitch.

(iii) **Quality or Timbre** Quality is the characteristic of sound that differentiates between two sounds of same intensity and same frequency.

Longitudinal Wave

There are three equations associated with any longitudinal wave

$$y\,(x, t),\ \Delta p\,(x, t) \text{ and } \Delta \rho\,(x, t)$$

y represents displacement of medium particles from their mean position parallel to direction of wave velocity.

From $y\,(x, t)$ equation, we can make $\Delta p\,(x, t)$ or $\Delta \rho\,(x, t)$ equations by using the fundamental relation between them,

$$\Delta p = -B \cdot \frac{\partial y}{\partial x}$$

and

$$\Delta \rho = -\rho \cdot \frac{\partial y}{\partial x}$$

$\Delta p_0 = BAk$ and $\Delta \rho_0 = \rho A k$

$\Delta p\,(x, t)$ and $\Delta \rho\,(x, t)$ are in same phase. But $y\,(x, t)$ equation has a phase difference of $\frac{\pi}{2}$ with other two equations.

Wave Speed

- Speed of transverse wave on a stretched wire,

$$v = \sqrt{\frac{T}{\mu}} = \sqrt{\frac{T}{\rho S}}$$

- Speed of longitudinal wave, $v = \sqrt{\frac{E}{\rho}}$

(i) In solids, $E = Y =$ Young's modulus of elasticity

$$\therefore \quad v = \sqrt{\frac{Y}{\rho}}$$

(ii) In liquids, $E = B =$ bulk modulus of elasticity

$$\therefore \quad v = \sqrt{\frac{B}{\rho}}$$

(iii) In gases, according to Newton,

$E = B_T =$ isothermal bulk modulus of elasticity $= p$

$$\therefore \quad v = \sqrt{\frac{p}{\rho}}$$

But results did not match with this formula.

Laplace made correction in it. According to him,

$E = B_S =$ adiabatic bulk modulus of elasticity $= \gamma p$

$$\therefore \quad v = \sqrt{\frac{\gamma p}{\rho}} = \sqrt{\frac{\gamma RT}{M}} = \sqrt{\frac{\gamma kT}{m}}$$

Effect of Temperature, Pressure and Relative Humidity in Speed of Sound in Air

- **With temperature** $v \propto \sqrt{T}$

$$\Rightarrow \qquad \frac{v_1}{v_2} = \sqrt{\frac{T_1}{T_2}}$$

If v_0 and v_t are velocities of sound in air at 0°C and t°C, then

$$v_t = v_0 \left(1 + \frac{t}{273}\right)^{1/2}$$

or
$$v_t = v_0 + 0.61\, t$$

- **With pressure** Pressure has no effect on speed of sound as long as temperature remains constant.
- **With relative humidity** With increase in relative humidity in air, density decreases. Hence, speed of sound increases.

Sound Level (L)

$$L = 10 \log_{10} \frac{I}{I_0} \qquad \text{(in dB)}$$

Here, I_0 = intensity of minimum audible sound = 10^{-12} Wm^{-2}.
On comparing loudness of two sounds we may write,

$$L_2 - L_1 = 10 \log_{10} \frac{I_2}{I_1}$$

In case of point source, $I \propto \frac{1}{r^2}$ or $\frac{I_2}{I_1} = \left(\frac{r_1}{r_2}\right)^2$.

In case of line source, $I \propto \frac{1}{r}$ or $\frac{I_2}{I_1} = \left(\frac{r_1}{r_2}\right)$.

Doppler Effect in Sound

- If v_s and v_o are the velocities along the line joining S and O, then

$$f' = f\left(\frac{v \pm v_m \pm v_o}{v \pm v_m \pm v_s}\right)$$

- If velocities v_s and v_o are along some other direction, the components of velocities along the line joining source and observer are taken.

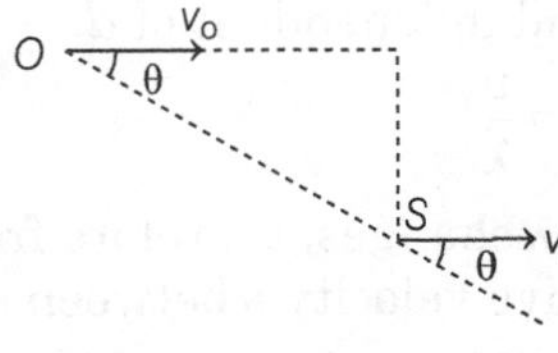

For example, in the figure shown,

$$f' = \left(\frac{v + v_o \cos\theta}{v + v_s \cos\theta}\right) f$$

- Change in frequency depends on the fact that whether the source is moving towards the observer or the observer is moving towards the source. But when the speed of source and observer are much lesser than that of sound, then the change in frequency becomes independent of the fact whether the source is moving or the observer.

  For example, suppose a source is moving towards a stationary observer with speed u and the speed of sound is v, then

$$f' = \left(\frac{v}{v-u}\right) f = \left(\frac{1}{1-\frac{u}{v}}\right) f$$

$$= \left(1 - \frac{u}{v}\right)^{-1} f$$

  Using the binomial expansion, we have

$$\left(1-\frac{u}{v}\right)^{-1} \approx 1 + \frac{u}{v} \qquad [\text{if } u << v]$$

$$\therefore \qquad f' = \left(1 + \frac{u}{v}\right) f$$

  On the other hand, if an observer moves towards a stationary source with same speed u, then

$$f' = \left(\frac{v+u}{v}\right) f = \left(1 + \frac{u}{v}\right) f$$

  which is same as above.

- As long as v_s and v_o are along the line joining S and O, so doppler's effect (or change in frequency) does not depend upon the distance between S and O. For example in the given figure,

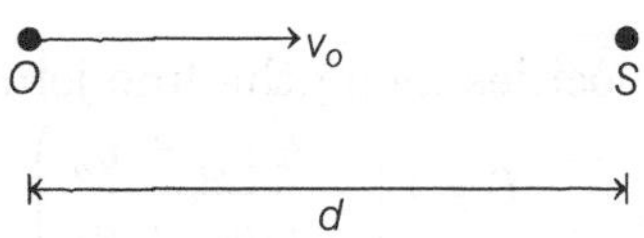

$$f' = f\left(\frac{v+v_o}{v}\right)$$

  $f' > f$ but it is constant and independent of d.

- Frequency is given by $f = \frac{v}{\lambda}$

  By the motion of source, λ changes, therefore frequency changes. From the motion of observer, relative velocity v between sound and observer changes, therefore frequency changes.

- Despite the motion of source or observer (or both), Doppler's effect is not observed (or $f' = f$) under the following four conditions.

***Condition* 1** $\mathbf{v}_s$ **or** $\mathbf{v}_o$ **is making an angle of 90° with the line joining** S **and** O. **This is illustrated in the following figure.**

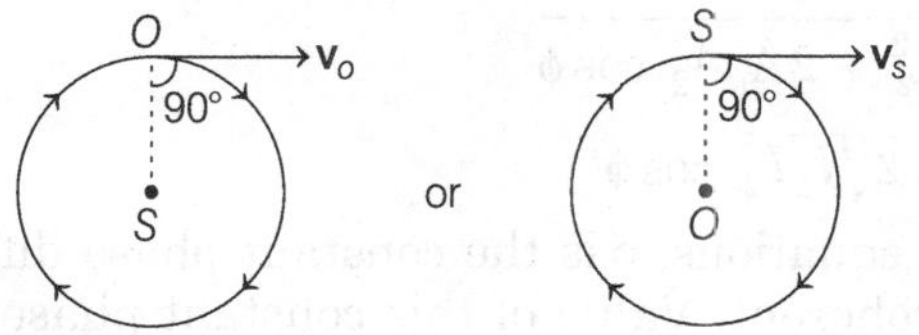

***Condition* 2** **Source and observer both are in motion but their velocities are equal or relative motion between them is zero.**

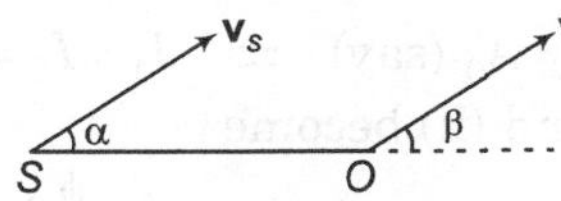

In the figure shown, $\mathbf{v}_s = \mathbf{v}_o$ **if** $v_s = v_o$ **and** $\alpha = \beta$.
Taking the components along SO, **we have**

$$f' = f\left(\frac{v - v_o \cos\beta}{v - v_s \cos\alpha}\right)$$

$\Rightarrow$ $f' = f$ **because** $v_s = v_o$ **and** $\alpha = \beta$

***Condition* 3** **Source and observer both are at rest. Only medium is in motion.**

In the figure shown, source and observer are at rest. Only wind is blowing in a direction from source to observer.

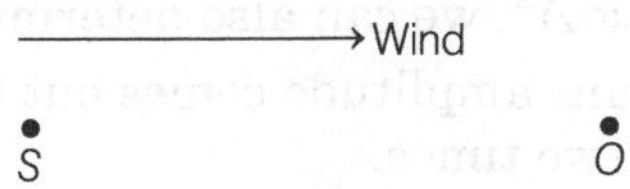

Then, change in frequency,

$$f' = f\left(\frac{v + v_{\text{wind}}}{v + v_{\text{wind}}}\right) = f$$

***Condition* 4** **A train is moving on a circular track. Engine is the source of sound and guard is the observer. Although, yet** f' **comes out to be** f.

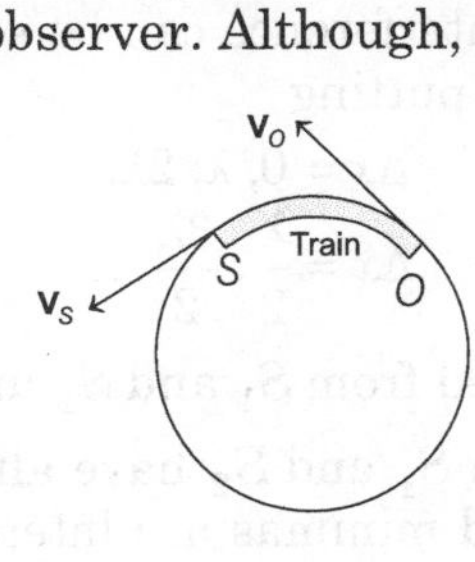

$\mathbf{v}_s \neq \mathbf{v}_o$

Principle of Superposition and Interference

- $y = y_1 + y_2$
- $A = \sqrt{A_1^2 + A_2^2 + 2A_1 A_2 \cos\phi}$ ...(i)
- $I = I_1 + I_2 + 2\sqrt{I_1 I_2} \cos\phi$ ...(ii)
- In the above equations, ϕ is the constant phase difference at that point, as the sources are coherent. Value of this constant phase difference will be different at different points.
- The special case of above two equations is, when the individual amplitudes (or intensities) are equal

  or $A_1 = A_2 = A_0$ (say) $\Rightarrow$ $I_1 = I_2 = I_0$ (say)

  In this case, Eqs. (i) and (ii) become

  $$A = 2A_0 \cos\frac{\phi}{2} \quad \text{...(iii)}$$

  and

  $$I = 4I_0 \cos^2\frac{\phi}{2} \quad \text{...(iv)}$$

- From Eqs. (i) to (iv), we can see that, for given values of A_1, A_2, I_1 and I_2 or the resultant amplitude and the resultant intensity are the functions of only ϕ.
- If three or more than three waves (due to coherent sources) meet at some point, then there is no direct formula for finding resultant amplitude or resultant intensity.

  In this case, first of all we will find resultant amplitude by vector method (either by using polygon law of vector addition or component method) and then by the relation $I \propto A^2$, we can also determine the resultant intensity.

  For example, if resultant amplitude comes out to be $\sqrt{2}$ times, then resultant intensity will become two times.
- In interference, two or more than two waves from coherent sources meet at several points. At different points, Δx, $\Delta\phi$ or ϕ, resultant amplitude and therefore resultant intensity will be different (varying from I_{max} to I_{min}).

  But whatever is the resultant intensity at some point, it remains constant at that point.
- Most of the problems of interference can be solved by calculating the net path difference Δx and then by putting

  $$\Delta x = 0, \lambda, 2\lambda, \ldots \quad \text{(For constructive interference)}$$

  $$\Delta x = \frac{\lambda}{2}, \frac{3\lambda}{2}, \ldots \quad \text{(For destructive interference)}$$

  provided the waves emitted from S_1 and S_2 are in phase.
- If two waves emitted from S_1 and S_2 have already a phase difference of π, the conditions of maximas and minimas are interchanged, i.e. path difference

  $$\Delta x = \frac{\lambda}{2}, \frac{3\lambda}{2} \ldots \quad \text{(For constructive interference)}$$

  and

  $$\Delta x = \lambda, 2\lambda, \ldots \quad \text{(For destructive interference)}$$

Reflection and Transmission of a Wave

Wave property	Reflection	Transmission (Refraction)
v	does not change	changes
f, T, ω	do not change	do not change
λ, k	do not change	change
A, I	change	change
ϕ	$\Delta\phi = 0$, from a rarer medium	$\Delta\phi = 0$
	$\Delta\phi = \pi$, from a denser medium	

- Amplitude in reflection as well as transmission, changes.

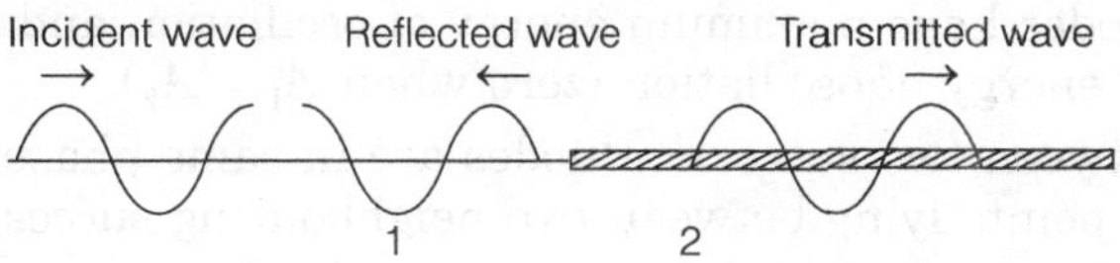

If amplitude of incident wave in medium-1 is A_i, it is partly reflected and partly transmitted at the boundary of two media-1 and 2. Wave speeds in two media are v_1 and v_2. If amplitudes of reflected and transmitted waves are A_r and A_t, then

$$A_r = \left(\frac{v_2 - v_1}{v_2 + v_1}\right) A_i$$

and
$$A_t = \left(\frac{2v_2}{v_1 + v_2}\right) A_i$$

From the above two expressions, we can make the following conclusions :

***Conclusion* 1** If $v_1 = v_2$, then $A_r = 0$ and $A_t = A_i$

Basically $v_1 = v_2$ means both media are same from wave point of view. So, in this case there is no reflection ($A_r = 0$), only transmission ($A_t = A_i$) is there.

***Conclusion* 2** If $v_2 < v_1$, then A_r comes out to be negative. Now, $v_2 < v_1$ means the second medium is denser. A_r in this case is negative means, there is a phase change of π.

***Conclusion* 3** If $v_2 > v_1$, then $A_t > A_i$. This implies that amplitude always increases as the wave travels from a denser medium to rarer medium (as $v_2 > v_1$, so second medium is rarer).

- **Power** At the boundary of two media,
energy incident per second = energy reflected per second
+ energy transmitted per second

or power incident = power reflected + power transmitted

or $$P_i = P_r + P_t$$

Stationary Waves

- Stationary waves are formed by the superposition of two identical waves travelling in opposite directions.
- Formation of stationary waves is really the interference of two waves in which coherent (same frequency) sources are required.
- By the word 'Identical waves' we mean that they must have same value of v, ω and k. Amplitudes may be different, but same amplitudes are preferred.
- In stationary waves, all particles oscillate with same value of ω but amplitudes vary from $A_1 + A_2$ to $A_1 - A_2$.

  Points where amplitude is maximum (or $A_1 + A_2$) are called antinodes (or points of constructive interference) and points where amplitude is minimum (or $A_1 - A_2$) are called nodes (or points of destructive interference).
- If $A_1 = A_2 = A$, then amplitude at antinode is $2A$ and at node is zero. In this case points at node do not oscillate.
- Points at antinodes have maximum energy of oscillation and points at nodes have minimum energy of oscillation (zero when $A_1 = A_2$).
- Points lying between two successive nodes are in same phase. They are out of phase with the points lying between two neighbouring successive nodes.
- Equation of stationary wave is of type,

  $$y = 2A \sin kx \cos \omega t$$

  or $$y = 2A \cos kx \sin \omega t, \text{ etc.}$$

  This equation can also be written as,

  $$y = A_x \sin \omega t \text{ or } y = A_x \cos \omega t$$

  If $x = 0$ is a node then, $A_x = A_0 \sin kx$

  If $x = 0$ is an antinode then, $A_x = A_0 \cos kx$

  Here, A_0 is maximum amplitude at antinode.
- Energy of oscillation in a given volume can be obtained either by adding energies due to two individual waves travelling in opposite directions or by integration. Because in standing wave amplitude and therefore energy of oscillation varies point to point.

Travelling waves	Stationary waves
In these waves, all particles of the medium oscillate with same frequency and amplitude.	In these waves, all particles except nodes (if $A_1 = A_2$) oscillate with same frequency but different amplitudes. Amplitudes is zero at nodes and maximum at antinodes
At any instant phase difference between any two particles can have any value between 0 and 2π.	At any instant phase difference between any two particles can be either zero or π.
In these waves, at no instant all the particles of the medium pass through their mean positions simultaneously.	In these waves, all particles of the medium pass through their mean positions simultaneously twice in each time period.
These waves transmit energy in the medium.	These waves do not transmit energy in the medium, provided $A_1 = A_2$.

Oscillations of Stretched Wire or Organ Pipes

- **Stretched wire**

Fundamental tone or first harmonic ($n = 1$)

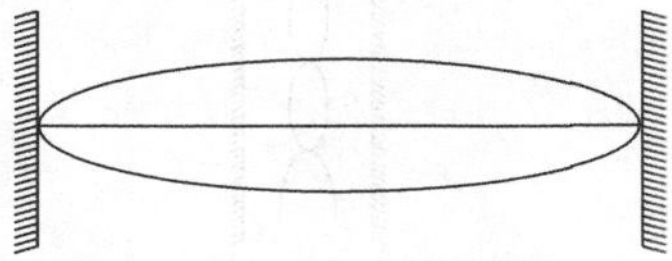

First overtone or second harmonic ($n = 2$)

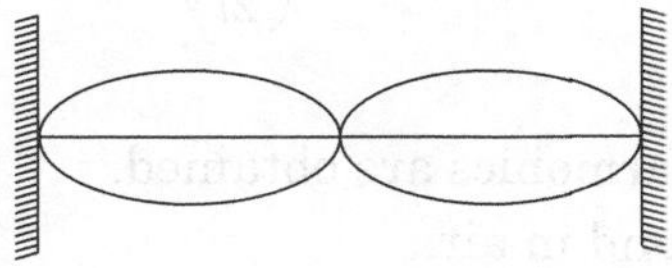

Second overtone or third harmonic ($n = 3$)

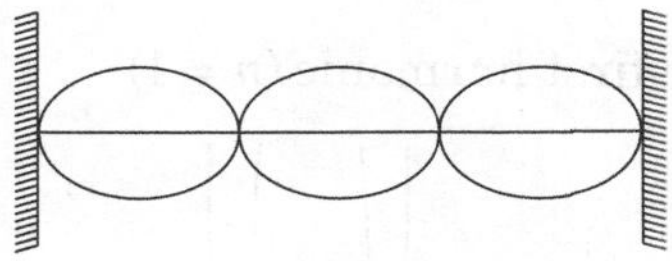

$$f = n\left(\frac{v}{2l}\right)$$

Here, $n = 1, 2, 3, \ldots$..

Even and odd both harmonics are obtained.

Here, $$v = \sqrt{\frac{T}{\mu}} \text{ or } \sqrt{\frac{T}{\rho S}}$$

- **Open organ pipe**

Fundamental tone or first harmonic ($n = 1$)

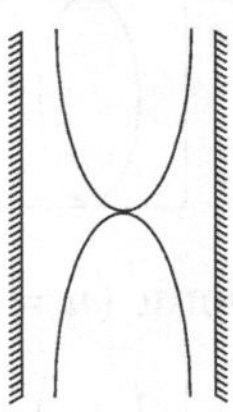

First overtone or second harmonic ($n = 2$)

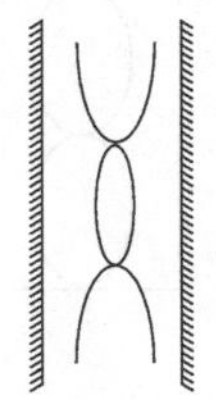

Second overtone or third harmonic (n = 3)

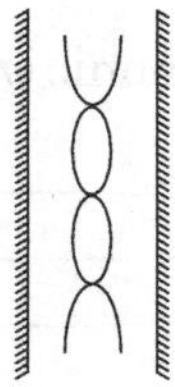

$$f = n\left(\frac{v}{2l}\right)$$

Here, n = 1, 2, 3, ...
Even and odd both harmonics are obtained.
Here, v = speed of sound in air.

v will be either given in the question, otherwise calculate from $v = \sqrt{\frac{\gamma RT}{M}}$.

- **Closed organ pipe**

Fundamental tone or first harmonic (n = 1)

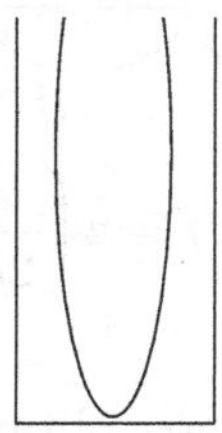

First overtone or third harmonic (n = 3)

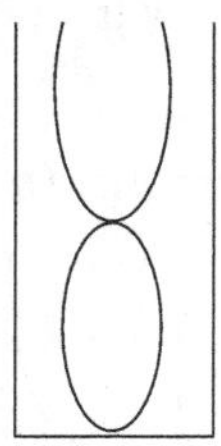

Second overtone or fifth harmonic (n = 5)

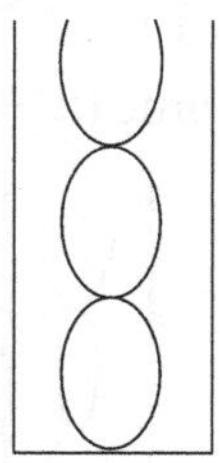

$$f = n\left(\frac{v}{4l}\right)$$

$$n = 1, 3, 5, ...$$

- Open end of pipe is displacement antinode, but pressure and density are nodes. Closed end of pipe is displacement node, but pressure and density are antinodes.
- Laplace correction $e = 0.6r$ (in closed pipe)

  and $2e = 1.2r$ (In open pipe)

  Hence, $f = n\left[\frac{v}{2(l + 1.2r)}\right]$ (In open pipe)

  with $n = 1, 2, 3......$

  and $f = n\left[\frac{v}{4(l + 0.6r)}\right]$ (In closed pipe)

  with $n = 1, 3, 5......$
- If an open pipe and a closed pipe are of same lengths, then fundamental frequency of open pipe is two times the fundamental frequency of closed pipe.

Note *Stationary transverse waves are formed in stretched wire and longitudinal stationary waves are formed in organ pipes.*

Beats Frequency

$$f_b = f_1 - f_2 \quad (f_1 > f_2)$$

The difference of frequencies should not be more than 10. Sound persists on human ear drums for 0.1 s. Hence, beats will not be heard if the frequency difference exceeds 10.

Resonance Tube

Resonance tube is a closed organ pipe in which length of air column can be changed by changing height of liquid column in it.

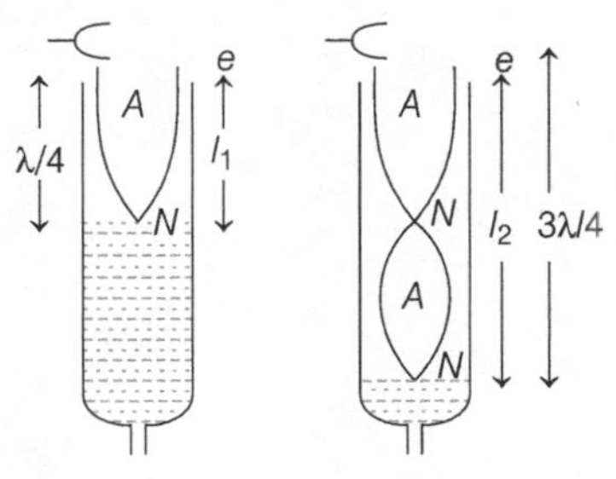

For first resonance, $\frac{\lambda}{4} = l_1 + e$

For second resonance, $\frac{3\lambda}{4} = l_2 + e$

$$\frac{3\lambda}{4} - \frac{\lambda}{4} = (l_2 - l_1)$$

$$\Rightarrow \qquad \frac{\lambda}{2} = (l_2 - l_1)$$

or
$$\lambda = 2(l_2 - l_1)$$

Velocity of sound, $v = f\lambda = 2f(l_2 - l_1)$

End correction, $e = \dfrac{l_2 - 3l_1}{2}$

Here, f = frequency of tuning fork

Echo

The repetition of sound caused by the reflection of sound waves is called an echo. Since, sound persists on ear for 0.1 s, so the minimum distance from a sound reflecting surface to hear an echo is 16.5 m.

CHAPTER 14

Heat and Thermodynamics

Different Scale of Temperature

(i) **Celsius Scale** In this scale of temperature, the melting point of ice is taken as 0°C and the boiling point of water as 100°C and the space between these two points is divided into 100 equal parts.

(ii) **Fahrenheit Scale** In this scale of temperature, the melting point of ice is taken as 32°F and the boiling point of water as 212°F and the space between these two points is divided into 180 equal parts.

(iii) **Kelvin Scale** In this scale of temperature, the melting point of ice is taken as 273 K and the boiling point of water as 373 K and the space between these two points is divided into 100 equal parts.

Relation between Different Scales of Temperatures

$$\frac{C}{100} = \frac{F - 32}{180} = \frac{K - 273}{100}$$

Calorimetry

- $Q = ms\Delta\theta = c\Delta\theta$, when temperature varies without change in state.

Note *If specific heat s is a function of temperature θ, then* $Q = \int_{\theta_i}^{\theta_f} ms d\theta$

- $Q = mL$, when state changes without change in temperature.
- s = specific heat of any substance
  = heat required to increase the temperature of unit mass by 1°C or 1 K.
- c = heat capacity of a body = ms
  = heat required to increase the temperature of whole body by 1°C or 1 K.

- Specific heat of water is 1 cal/g-°C between 14.5°C and 15.5°C.
- L = latent heat of any substance
  = heat required to convert unit mass of that substance from one state to another state.
- **Water equivalent of a vessel** It is the mass of equivalent water which takes same amount of heat as taken by the vessel for same rise of temperature.

Thermal Expansion

- $\Delta l = l \alpha \Delta\theta$, $\Delta s = s \beta \Delta\theta$ and $\Delta V = V \gamma \Delta\theta$
- $\beta = 2\alpha$ and $\gamma = 3\alpha$ for isotropic medium.
- **Anomalous expansion of water** When temperature of water is increased from 0°C, then its volume decreases upto 4°C, becomes minimum at 4°C and then increases.

  So, volume of water at 4°C is minimum and density is maximum. This behaviour of water around 4°C is called, anomalous expansion of water.

Effect of Temperature on Different Physical Quantities

- **On density** With increase in temperature, volume of any substance increases while mass remains constant, therefore density should decrease.

$$\rho' = \frac{\rho}{1 + \gamma\Delta\theta}$$

or $$\rho' \approx \rho(1 - \gamma \cdot \Delta\theta), \text{ if } \gamma \cdot \Delta\theta << 1$$

- **In fluid mechanics**

  ***Case* 1** When a solid whose density is less than the density of liquid is floating, then a fraction of its volume remains immersed in liquid. This fraction is

$$f = \frac{\rho_s}{\rho_l}$$

When temperature is increased, ρ_s and ρ_l both will decrease. Hence, fraction may increase, decrease or remain same. At higher temperature,

$$f' = f\left(\frac{1 + \gamma_l \Delta\theta}{1 + \gamma_s \Delta\theta}\right)$$

If $\gamma_l > \gamma_s$, $f' > f$ or immersed fraction will increase.

***Case* 2** When a solid whose density is more than the density of liquid is immersed completely in a liquid, then upthrust will act on 100% volume of solid and apparent weight appears less than the actual weight.

$$w_{\text{app}} = w - F$$

Here, $$F = V_s \rho_l g$$

With increase in temperature, V_s will increase and ρ_l will decrease, while g will remain unchanged.

Therefore, upthrust may increase, decrease or remain same.

At some higher temperature,

$$F' = F\left(\frac{1+\gamma_s\,\Delta\theta}{1+\gamma_l\Delta\theta}\right)$$

If $\gamma_s > \gamma_l$, upthrust will increase. Therefore, apparent weight will decrease.

- **Time period of pendulum**

$$T = 2\pi\sqrt{\frac{l}{g}} \quad \text{or} \quad T \propto \sqrt{l}$$

With increase in temperature, length of pendulum will increase. Therefore, time period will increase and pendulum clock will become slow and it loses the time. At some higher temperature,

$$T' = T\,(1+\alpha\,\Delta\theta)^{\frac{1}{2}}$$

or

$$T' \approx T\left(1+\frac{1}{2}\,\alpha\Delta\theta\right), \text{ if } \alpha\,\Delta\,\theta << 1$$

$$\Delta T = (T' - T) = \frac{1}{2}\,T\,\alpha\,\Delta\theta$$

Time lost/gained

$$\Delta t = \frac{\Delta T}{T'} \times t$$

- **Thermal stress** If temperature of a rod fixed at both ends is increased, then thermal stresses are developed in the rod.

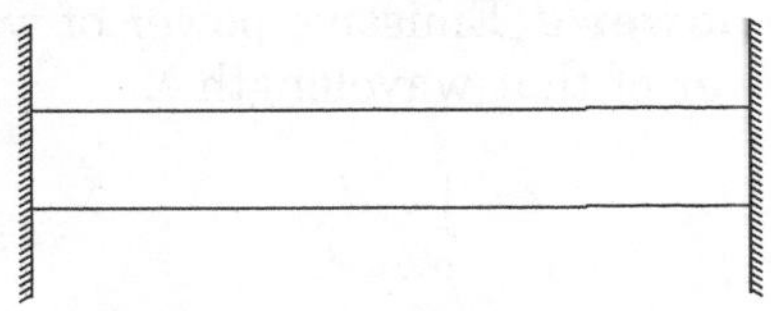

At some higher temperature, natural length of the rod should be more (by $\Delta l = l\alpha\Delta\theta$). We may assume that the rod has been compressed by a length,

$$\Delta l = l\,\alpha\,\Delta\theta$$

or

$$\text{strain} = \frac{\Delta l}{l} = \alpha\,\Delta\theta$$

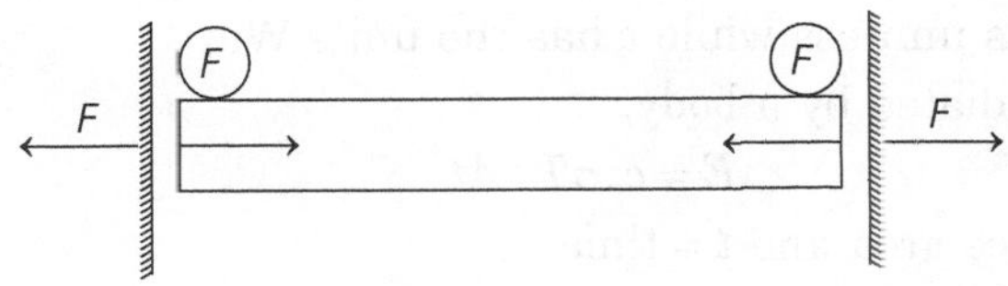

stress = $Y \times$ strain = $Y\,\alpha\,\Delta\theta$

where, Y = Young's modulus of elasticity.

$\therefore \quad F = A \times \text{stress} = YA\alpha\Delta\theta$

Rod applies this much force on wall to expand. In turn, wall also exerts equal and opposite pair of encircled forces on rod. Due to this pair of forces only, we can say that rod is compressed.

Heat Transfer, Heat Conduction through a Rod

- Heat flow in steady state $Q = \frac{KA(\theta_1 - \theta_2)}{l} t$
- Rate of flow of heat = heat current

or $$H = \frac{dQ}{dt} = \frac{\text{TD}}{R}$$

Here, TD = temperature difference = $\theta_1 - \theta_2$

and R = thermal resistance = $\frac{l}{KA}$

Radiation

- Absorptive power $a = \frac{\text{energy absorbed}}{\text{energy incident}}$

$$a \leq 1$$

$a = 1$ for perfectly black body.

- Spectral absorptive power a_λ = absorptive power of wavelength λ

$$a_\lambda \leq 1$$

$a_\lambda = 1$ for perfectly black body.

- **Emissive power *e*** Energy radiated per unit area per unit time is called emissive power of a body. Its SI units are $\text{Js}^{-1}\text{m}^{-2}$ or Wm^{-2}.
- **Spectral emissive power e_λ** Emissive power of wavelength λ is known as spectral emissive power of that wavelength λ.

$$e = \int_0^\infty e_\lambda d_\lambda$$

- **Stefan's law** Emissive power of a body is given by

$$e = e_r \sigma T^4$$

Here, e_r = emissivity, emittance, relative emissivity or relative emittance.

$e_r \leq 1$

$e_r = 1$ for a perfectly black body.

Sometimes emissivity is also denoted by *e*. In that case, differentiate them by their units. e_r is unitless while *e* has the units Wm^{-2}.

- Total energy radiated by a body,

$$E = e_r \sigma T^4 At$$

Here, A = surface area and t = time.

Also, $a = e_r$ or absorptivity of a body = its emissivity.

- **Kirchhoff's law** If different bodies (including a perfectly black body) are kept at same temperature, then

$$e_\lambda \propto a_\lambda \text{ or } \frac{e_\lambda}{a_\lambda} = \text{constant}$$

or $$\left(\frac{e_\lambda}{a_\lambda}\right)_{\text{body-1}} = \left(\frac{e_\lambda}{a_\lambda}\right)_{\text{body-2}}$$

$$= \left(\frac{e_\lambda}{a_\lambda}\right)_{\text{perfectly black body}}$$

$$= (e_\lambda)_{\text{perfectly black body}}$$

From this law, following two conclusions can be drawn

(i) Good absorbers of a particular wavelength λ are also good emitters of same wavelength λ.

(ii) At a given temperature, ratio of e_λ and a_λ for any body is constant. This ratio is equal to e_λ of perfectly black body at that temperature.

- **Wien's displacement law**

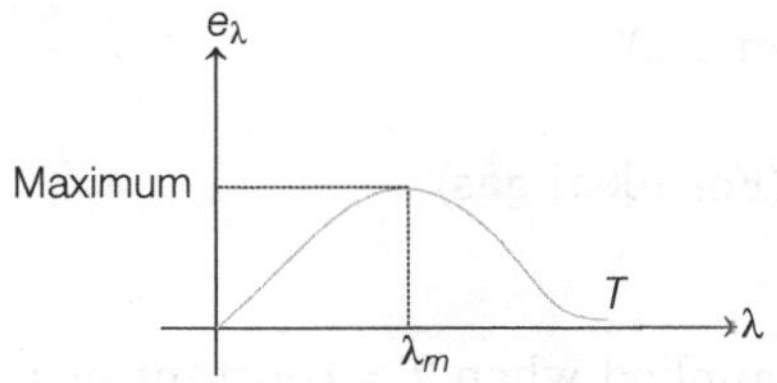

$$\lambda_m \propto \frac{1}{T} \text{ or } \lambda_m T = \text{constant} = \text{Wien's constant } b$$

Here, $b = 2.89 \times 10^{-3}$ m-K

Further, area of this graph will give total emissive power which is proportional to T^4.

- **Cooling of a body by radiation**

(i) Rate of cooling

$$-\frac{dT}{dt} = \frac{e_r A\sigma}{ms}(T^4 - T_0^4) \quad \text{or} \quad -\frac{dT}{dt} \propto T^4 - T_0^4$$

(ii) **Newton's law of cooling** If temperature difference of a body with atmosphere is small, then rate of cooling ∝ temperature difference.

(iii) If body cools by radiation according to Newton, then temperature of body decreases exponentially.

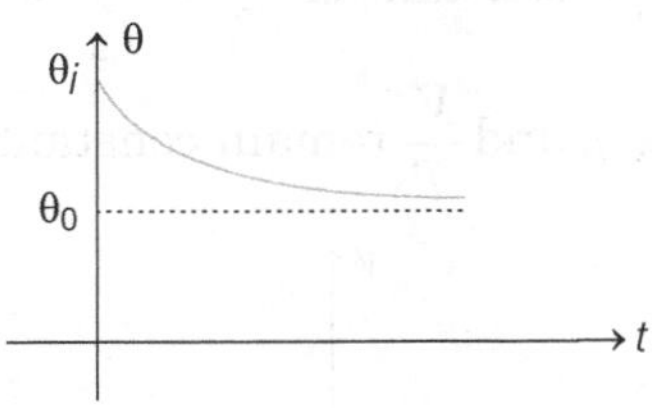

In the figure,

θ_i = initial temperature of body

θ_0 = temperature of atmosphere.

Temperature at any time t can be written as

$$\theta = \theta_0 + (\theta_i - \theta_0)\, e^{-\alpha t}$$

(iv) According to Newton, if body is cooling, then to find temperature of a body at any time t, we will have to calculate $e^{-\alpha t}$. To avoid this, you can use a shortcut approximate formula given below

$$\left(\frac{\theta_1 - \theta_2}{t}\right) = \alpha\left[\frac{\theta_1 + \theta_2}{2} - \theta_0\right].$$

Kinetic Theory of Gases

Different equations used in kinetic theory of gases are listed below

- $pV = nRT = \dfrac{m}{M}RT$ (where, m = mass of gas)
- Density, $\rho = \dfrac{m}{V}$ (General)

  $= \dfrac{pM}{RT}$ (For ideal gas)
- **Gas laws**

  (i) **Boyle's law** is applied when T = constant or process is isothermal. In this condition,

  $$pV = \text{constant or } p_1V_1 = p_2V_2 \text{ or } p \propto \frac{1}{V}$$

  In isothermal process, T, pV and U remain constant

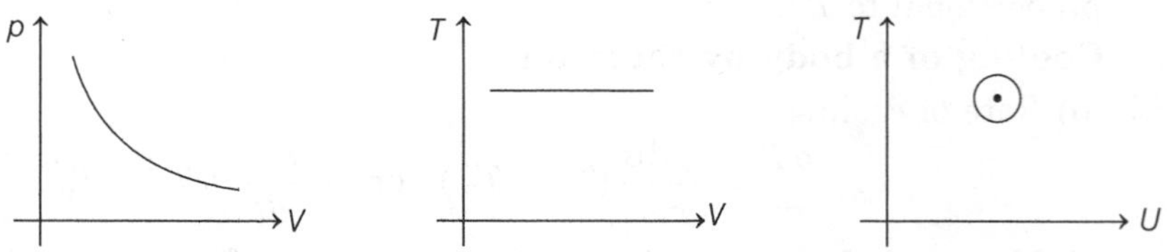

  (ii) **Charles' law** is applied when p = constant or process is isobaric. In this condition,

  $$\frac{V}{T} = \text{constant or } \frac{V_1}{T_1} = \frac{V_2}{T_2} \text{ or } V \propto T$$

  In isobaric process, p and $\dfrac{V}{T}$ remain constant.

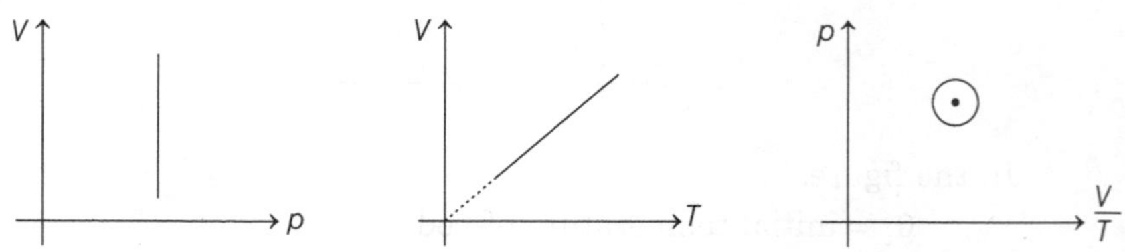

  (iii) **Pressure law or Gay-Lussac's law** is applied when V = constant or process is isochoric.

In this condition,

$$\frac{p}{T} = \text{constant}$$

or $$\frac{p_1}{T_1} = \frac{p_2}{T_2} \quad \text{or} \quad p \propto T$$

In isochoric process V, $\frac{p}{T}$ and ρ remain constant.

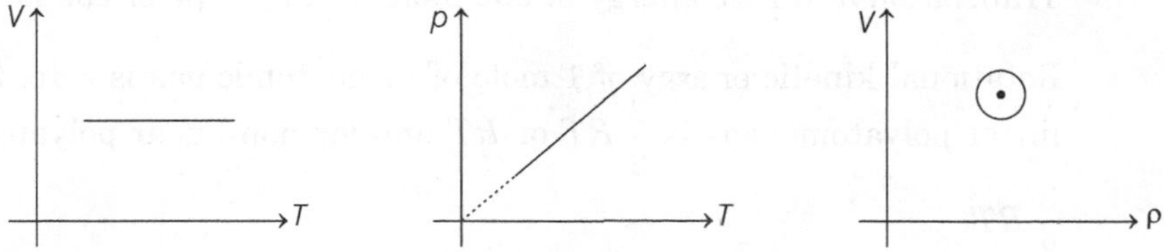

- For speeds, $v = \sqrt{\frac{ART}{M}} = \sqrt{\frac{AkT}{m}} = \sqrt{\frac{Ap}{\rho}}$

  Here, m = mass of one gas molecule.

  $A = 3$, for rms speed of gas molecules

  $A = \frac{8}{\pi} \approx 2.5$ for average speed of gas molecules

  $A = 2$, for most probable speed of gas molecules

- $p = \frac{1}{3}\frac{mn}{V} v_{rms}^2$

  Here, m is the mass of one gas molecule and n is total number of molecules.

- $p = \frac{2}{3} E.$

  Here, E = total translational kinetic energy per unit volume

- f = degree of freedom

  = 3 for monoatomic gas

  = 5 for diatomic and linear polyatomic gas

  = 6 for non-linear polyatomic gas

Note *(i) Vibrational degree of freedom is not taken into consideration.*
(ii) Translational degree of freedom for any type of gas is three.

- Total internal energy of a gas is

$$U = \frac{nf}{2} RT$$

  Here, n = total number of gram moles.

- $C_V = \frac{dU}{dT}$

  where, U = internal energy of one mole of a gas $= \frac{f}{2} RT$

  $\therefore \quad C_V = \frac{f}{2} R = \frac{R}{\gamma - 1}$

- $C_p = C_V + R = \left(1 + \frac{f}{2}\right) R = \left(\frac{\gamma}{\gamma - 1}\right) R$
- $\gamma = \frac{C_p}{C_V} = 1 + \frac{2}{f}$
- Internal energy of 1 mole in one degree of freedom for any gas is $\frac{1}{2} RT$.
- Translational kinetic energy of one mole for any type of gas is $\frac{3}{2} RT$.
- Rotational kinetic energy of 1 mole of monoatomic gas is zero, for diatomic or linear polyatomic gas is $\frac{2}{2} RT$ or RT and for non-linear polyatomic gas is $\frac{3}{2} RT$.
-

Nature of gas	Degree of freedom			Internal energy of 1 mole			Internal energy of 1 molecule		
	Total	Translational	Rotational	Total	Translational	Rotational	Total	Translational	Rotational
Monoatomic	3	3	0	$\frac{3}{2}RT$	$\frac{3}{2}RT$	0	$\frac{3}{2}kT$	$\frac{3}{2}kT$	0
Dia or linear polyatomic	5	3	2	$\frac{5}{2}RT$	$\frac{3}{2}RT$	RT	$\frac{5}{2}kT$	$\frac{3}{2}kT$	kT
Non-linear polyatomic	6	3	3	$3RT$	$\frac{3}{2}RT$	$\frac{3}{2}RT$	$3kT$	$\frac{3}{2}kT$	$\frac{3}{2}kT$

-

Nature of gas	f	$U = \frac{f}{2}RT$	$C_V = dU/dT = \frac{f}{2}R$	$C_p = C_V + R$	$\gamma = \frac{C_p}{C_V} = 1 + \frac{2}{f}$
Monoatomic	3	$\frac{3}{2}RT$	$\frac{3}{2}R$	$\frac{5}{2}R$	$\frac{5}{3} = 1.67$
Dia and linear polyatomic	5	$\frac{5}{2}RT$	$\frac{5}{2}R$	$\frac{7}{2}R$	$\frac{7}{5} = 1.4$
Non-linear polyatomic	6	$3RT$	$3R$	$4R$	$\frac{4}{3} = 1.33$

- **Mixture of non-reactive gases**

(i) $n = n_1 + n_2$

(ii) $p = p_1 + p_2$

(iii) $U = U_1 + U_2$

(iv) $\Delta U = \Delta U_1 + \Delta U_2$

(v) $C_V = \frac{n_1 C_{V_1} + n_2 C_{V_2}}{n_1 + n_2}$

(vi) $C_p = \frac{n_1 C_{p_1} + n_2 C_{p_2}}{n_1 + n_2} = C_V + R$

(vii) $\gamma = \frac{C_p}{C_V}$ or $\frac{n}{\gamma - 1} = \frac{n_1}{\gamma_1 - 1} + \frac{n_2}{\gamma_2 - 1}$

(viii) $M = \frac{n_1 M_1 + n_2 M_2}{n_1 + n_2}$

- **Mean free path** Every gas consists of a very large number of molecules. These molecules are in a state of continuous rapid and random motion. They undergo perfectly elastic collisions against one another.

Therefore, path of a single gas molecule consists of a series of short *zig-zag* paths of different lengths.

The mean free path of a gas molecule is the average distance between two successive collisions. It is represented by λ.

$$\lambda = \frac{kT}{\sqrt{2}\pi\sigma^2 p}$$

Here, σ = diameter of the molecule
and k = Boltzmann's constant.

- **Avogadro's hypothesis** At constant temperature and pressure, equal volumes of different gases contain equal number of molecules. In 1 g-mol of any gas, there are 6.02×10^{23} molecules of that gas. This is called Avogadro's number. Thus,

$$N = 6.02 \times 10^{23} \text{ per g-mol}$$

Therefore, the number of molecules in mass m of the substance

$$= nN = \frac{m}{M} \times N$$

- **Dalton' law of partial pressure** According to this law, if the gases filled in a vessel do not react chemically, then the combined pressure of all the gases is due to the partial pressure of the molecules of the individual gases.

If $p_1, p_2, \ldots$ represent the partial pressures of the different gases, then the total pressure is,

$$p = p_1 + p_2 \ldots$$

Thermodynamics

- **Molar heat capacity** = Heat required to raise the temperature of 1 mole of any substance by 1°C or 1 K.

$$C = \frac{Q}{n\Delta T}$$

$$\Rightarrow \quad Q = nC\Delta T$$

Molar heat capacity of solids and liquids is almost constant.

In case of gases, C is process dependent. It varies from 0 to ∞.

In isothermal process, $C = \infty$ as $\Delta T = 0$

In adiabatic process, $C = 0$ as $Q = 0$

C_p (molar heat capacity of isobaric process) and C_V (molar heat capacity of isochoric process) are commonly used.

In a general process pV^x = constant, molar heat capacity is given by

$$C = \frac{R}{\gamma - 1} + \frac{R}{1 - x}$$

- **First law of thermodynamics** This is the law of conservation of energy is given by

$$Q = \Delta U + W$$

This law can be applied for any type of system (solid, liquid or gas) but in most of the cases, our system is an ideal gas.

- **Detailed discussion of three terms of first law of thermodynamics**

(i) **Work done** Following methods are generally used to find the work done.

Method 1 $W = \int_{V_i}^{V_f} p dV$ (because $dW = pdV$)

Here, p should be either constant or function of V. If p is constant, it means process is isobaric, $W = p(V_f - V_i) = p\Delta V$.

Method 2 Work done can also be obtained by area under p-V diagram with projection on V-axis.

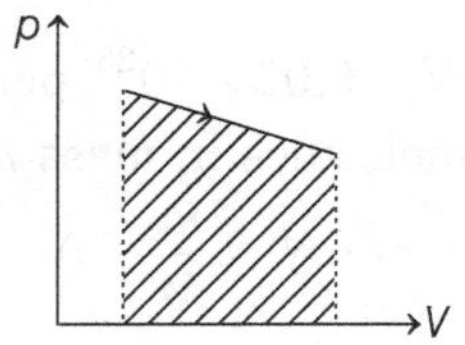

$W = +$ ve as volume is increasing

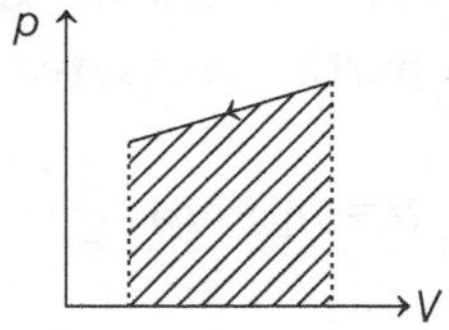

$W = -$ ve as volume is decreasing

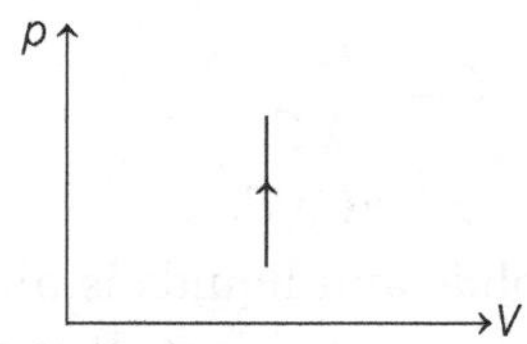

$W = 0$ as volume is constant

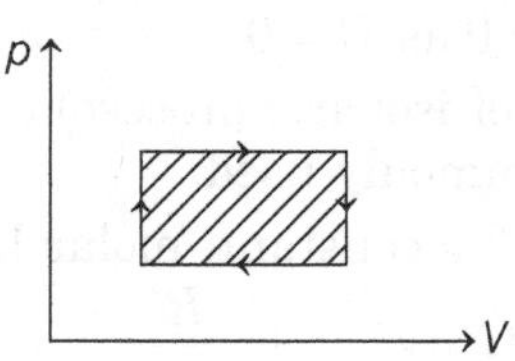

$W = +$ve as cyclic process is clockwise with p on V-axis.

(ii) **Change in internal energy** $\Delta U = nC_V \Delta T$ for all processes. For this, C_V (or nature of gas), n and ΔT should be known. If either of the three terms is not known, we can calculate ΔU by

$$\Delta U = Q - W$$

(iii) **Heat exchange** $Q = nC\Delta T$. For this, n, ΔT and molar heat capacity C should be known. C depends upon nature of gas and process. If either of the three terms (n, ΔT or C) is not known, we can calculate Q by,

$$Q = \Delta U + W$$

Name of the process	Important points in the process	$Q = nC\Delta T = W + \Delta U$	$\Delta U = nC_V \Delta T$	W
Isothermal	T, pV, U = constant $\Delta T = \Delta(pV) = \Delta U = 0$ $p_1V_1 = p_2V_2$ or $p \propto \frac{1}{V}$	$Q = W$	0	$nRT \ln\left(\frac{V_f}{V_i}\right)$ $= nRT \ln\left(\frac{p_i}{p_f}\right)$
Isochoric	$V, p, \frac{p}{T}$ = constant $\Delta V = \Delta p = \Delta\left(\frac{p}{T}\right) = 0$ $\frac{p_1}{T_1} = \frac{p_2}{T_2}$ or $p \propto T$	$C = C_V$ $\therefore\ Q = nC_V\Delta T$	$nC_V\Delta T$	0
Isobaric	$p, \frac{V}{T}$ = constant $\Delta p = \Delta\left(\frac{V}{T}\right) = 0$ $\frac{V_1}{T_1} = \frac{V_2}{T_2}$ or $V \propto T$	$C = C_p$ $\therefore\ Q = nC_p\Delta T$	$nC_V\Delta T$	$p\Delta V \rightarrow$ For any system $Q - \Delta U = n(C_p - C_V)\Delta T$ $= nR\Delta T \rightarrow$ for an ideal gas
Adiabatic process	pV^γ = constant $TV^{\gamma-1}$ = constant $T^\gamma p^{1-\gamma}$ = constant	0	$nC_V\Delta T$	$W = -\Delta U$ $\therefore\ W = -nC_V\Delta T$ $= -n\left(\frac{R}{\gamma-1}\right)(T_f - T_i)$ $= \frac{p_iV_i - p_fV_f}{\gamma - 1}$
Cyclic process	$(p_i, V_i, T_i) = (p_f, V_f, T_f)$ Since, $T_i = T_f$ $\Rightarrow\ U_i = U_f$ or $\Delta T = \Delta U = 0$	$Q_{net} = W_{net}$	0	W_{net} = area between cycle on p-V diagram
Polytropic process pV^X = constant	$C = \frac{R}{\gamma-1} + \frac{R}{1-x}$ $= C_V + \frac{R}{1-x}$	$nC\Delta T$	$nC_V\Delta T$	$Q - \Delta U = \frac{nR\Delta T}{1-x}$ $= \frac{nR(T_i - T_f)}{1-x}$ $= \frac{(p_iV_i - p_fV_f)}{1-x}$
Free expansion in vacuum	$\Delta U = 0$ $\Rightarrow\ U, T$ and pV = constant or $p \propto \frac{1}{V}$	0	0	0

- $\frac{dp}{dV} = -\frac{xP}{V}$ or pV^x = constant

or slope of p-V graph $= -x\dfrac{p}{V}$

In isobaric process $x = 0$, therefore slope $= 0$

In isothermal process $x = 1$, therefore slope $= -\dfrac{p}{V}$

In adiabatic process $x = \gamma$, therefore slope $= -\gamma\dfrac{p}{V}$

Adiabatic and isothermal expansion of an ideal gas

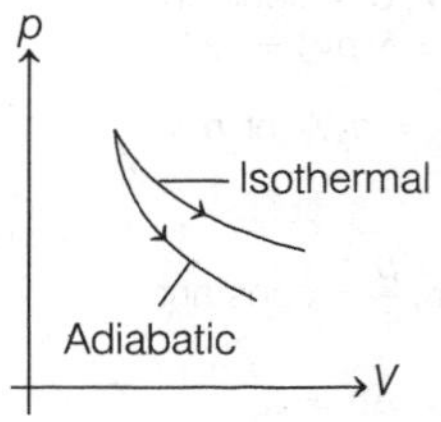

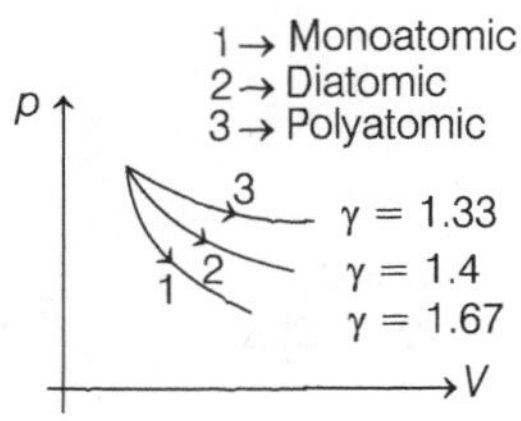

Adiabatic expansion of mono,dia and polyatomic gases.

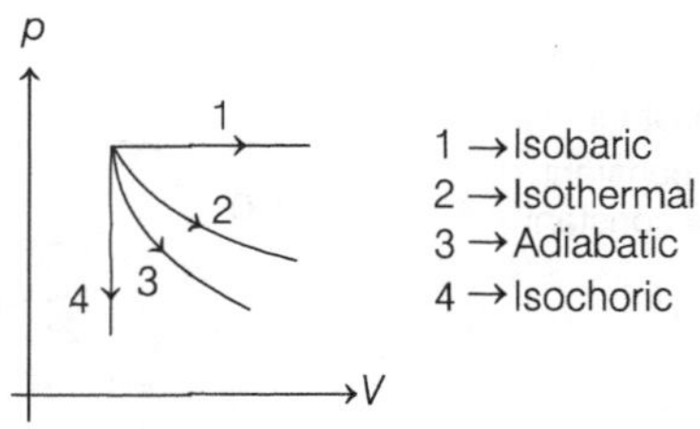

Heat Engines

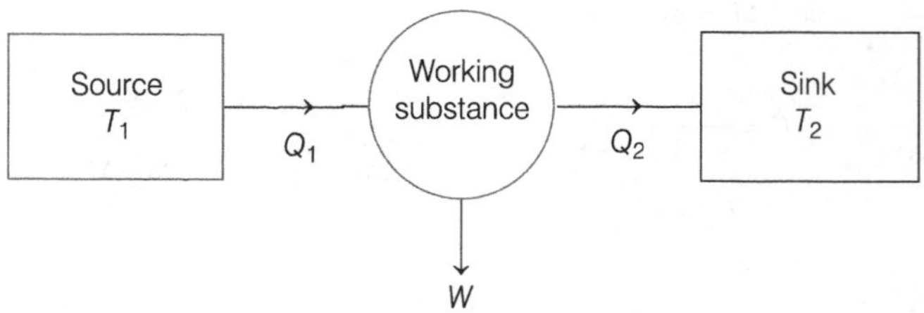

In general, $\eta = \dfrac{W}{Q_1} = \dfrac{Q_1 - Q_2}{Q_1} = 1 - \dfrac{Q_2}{Q_1}$

or $$\eta = \left(\frac{\text{Work done by the working substance during a cycle}}{\text{Heat supplied to the gas during the cycle}}\right) \times 100$$

$$= \frac{W_{\text{Total}}}{|Q_{+\text{ve}}|} \times 100 = \frac{|Q_{+\text{ve}}| - |Q_{-\text{ve}}|}{|Q_{+\text{ve}}|} \times 100$$

$$= \left\{1 - \left|\frac{Q_{-\text{ve}}}{Q_{+\text{ve}}}\right|\right\} \times 100$$

Thus, $$\eta = \frac{W_{\text{total}}}{|Q_{+\text{ve}}|} \times 100 = \left\{1 - \left|\frac{Q_{-\text{ve}}}{Q_{+\text{ve}}}\right|\right\} \times 100$$

Note *There cannot be a heat engine whose efficiency is 100%. It is always less than 100%. Thus,*

$$\eta \neq 100\%$$

or $W \neq Q_1$ *or* $W_{net} \neq Q_{+ve}$

or $Q_2 \neq 0$ *or* $|Q_{-ve}| \neq 0$

Carnot Engine

- Carnot cycle consists of the following four processes :
  (i) Isothermal expansion (process AB) at source temperature T_1
  (ii) Adiabatic expansion (process BC)
  (iii) Isothermal compression (process CD) at sink temperature T_2 and
  (iv) Adiabatic compression (process DA)

  The p-V diagram of the cycle is shown in the figure.

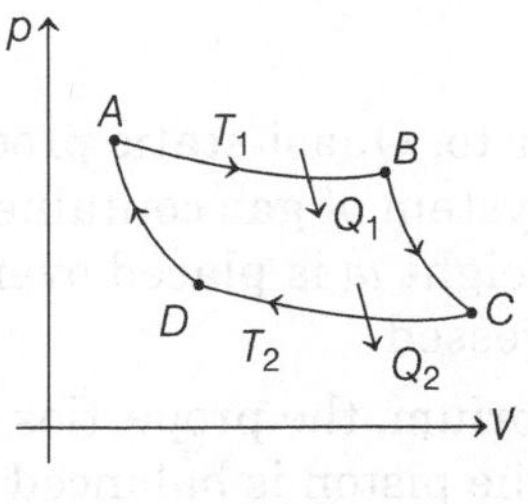

- In the whole cycle only Q_1 is the positive heat and Q_2 the negative heat. Thus,

$$Q_{+ve} = Q_1 \quad \text{and} \quad |Q_{-ve}| = Q_2$$

$$\therefore \quad \eta = \left(1 - \frac{Q_2}{Q_1}\right) \times 100$$

Specially for Carnot cycle, $\frac{Q_2}{Q_1}$ also comes out to be $\frac{T_2}{T_1}$.

$$\therefore \qquad \eta = \left(1 - \frac{T_2}{T_1}\right) \times 100$$

- Efficiency of Carnot engine is maximum (not 100%) for given temperatures T_1 and T_2. But still Carnot engine is not a practical engine because many ideal situations have been assumed while designing this engine which can practically not be obtained.

Adiabatic and Diathermic Wall

- **Adiabatic wall** An insulating wall (can be movable also) that does not allow flow of energy (heat) from one chamber to another is called an adiabatic wall.

  If two thermodynamic systems A and B are separated by an adiabatic wall, then the thermodynamic state of A will be independent of the state of B and *vice-versa*, if wall is fixed or if the wall is movable, then only pressure will be same on both sides.

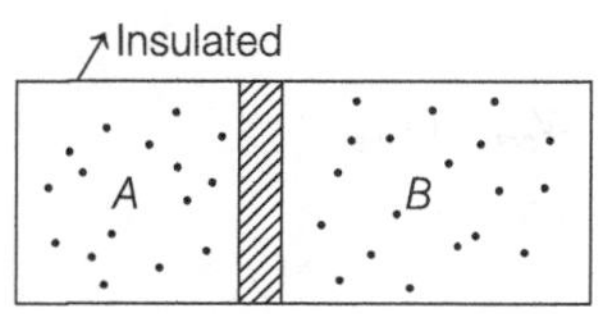

- **Diathermic wall** A conducting wall that allows energy flow (heat) from one chamber to another is called a diathermic wall. If two thermodynamic systems A and B are separated by a diathermic wall, then thermal equilibrium (same temperature) is attained in due course of time (if wall is fixed).

  If wall is movable, then temperature and pressure on both sides will become same.

Note *In the above two cases, thermodynamic systems A and B are insulated from the external surroundings.*

Quasi-Static Process

Quasi means almost or near to. Quasi-static process means very nearly static process. Let us consider a system of gas contained in a cylinder. The gas is held by a moving piston and a weight w is placed over the piston. Due to the weight , the gas in cylinder is compressed.

After the gas reaches equilibrium, the properties of gas are denoted by p_1, V_1 and T_1. The weight placed over the piston is balanced by upward force exerted by the gas. If the weight is suddenly removed, then there will be an unbalanced force between the system and the surroundings. The gas under pressure will expand and push the piston upwards.

The properties at this state after reaching equilibrium are p_2, V_2 and T_2. But the intermediate states passed through, when the system was in non-equilibrium states which cannot be described by thermodynamic coordinates. In this case, we only have initial and final states and do not have a path connecting them.

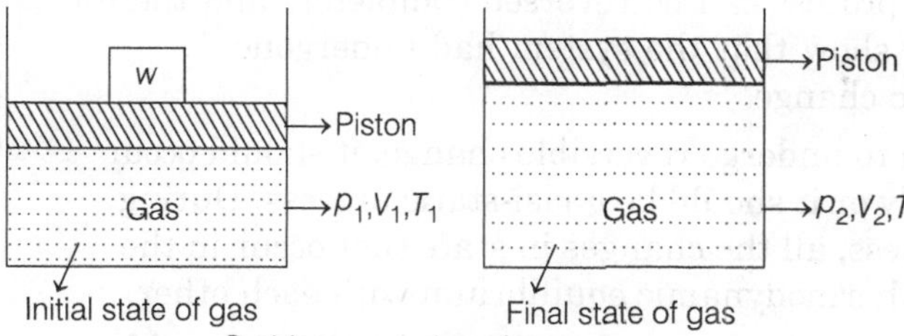

Sudden expansion of gas

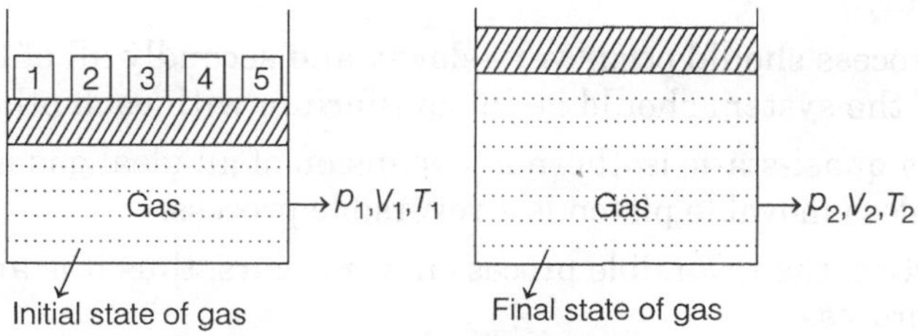

Quasi-static expansion of gas

Suppose, the weight is made of large numbers of small weights and one by one each of these small weights are removed and allowed the system to reach an equilibrium state. Then, we have intermediate equilibrium states and the path described by these states will not deviate much from the thermodynamic equilibrium state.

Such a process, which is the locus of all the intermediate points passed by the system is known as quasi-static process.

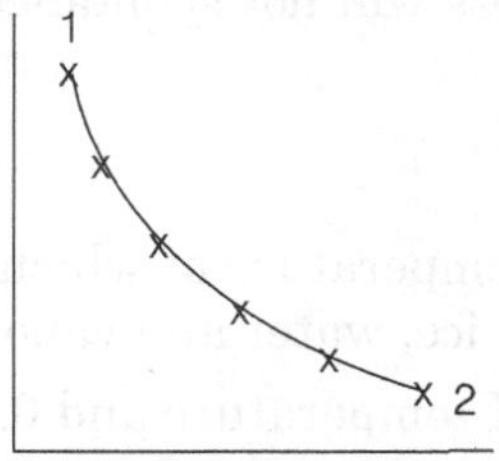

A quasi-static process

It means, this process is almost near to the thermodynamically equilibrium process. Infinite slowness is the characteristic feature of quasi-static process.

Reversible and Irreversible Process

- **Reversible process** **The process in which the system and surroundings can be restored to the initial state from the final state without producing any changes in the thermodynamic properties of the universe is called a reversible process.**
- **In the figure below, let us suppose that the system has undergone a change from state A to state B. If the system can be restored from state B to state A and there is no change in the universe, then the process is said to be a reversible process.**

- The reversible process can be reversed completely and there is no trace left to show that the system had undergone thermodynamic change.

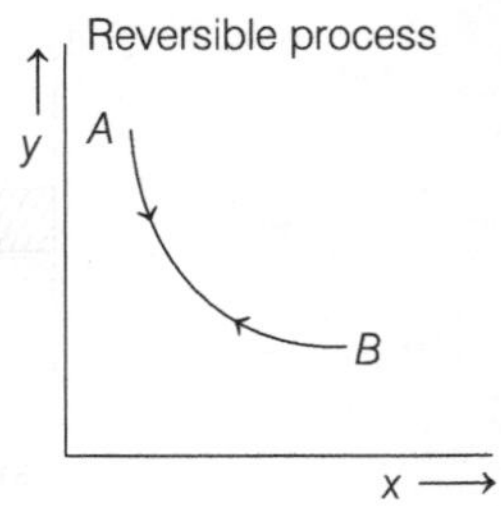

For the system to undergo reversible change, it should occur infinitely slowly or it should be quasi-static process. During reversible process, all the changes in state that occur in the system are in thermodynamic equilibrium with each other.

Thus, there are two important conditions for the reversible process to occur.

Firstly, the process should occur very slowly and secondly all of the initial and final states of the system should be in equilibrium with each other.

For example, a quasi-static isothermal expansion of an ideal gas in a cylinder fitted with a frictionless movable piston is a reversible process.

In actual practice, the reversible process never occurs, thus it is an ideal or hypothetical process.

- **Irreversible process** The process is said to be an irreversible process, if it cannot return the system and the surroundings to their original conditions when the process is reversed.

The irreversible process is not at equilibrium throughout the process. Several examples can be cited. For examples,

(i) When we are driving the car uphill, it consumes a lot of fuel and this fuel is not returned when we are driving down the hill.

(ii) Cooking gas leaking from a gas cylinder in the kitchen diffuses to the entire room. The diffusion process will not spontaneously reverse and bring the gas back to the cylinder.

Triple Point of Water

The values of pressure and temperature at which water coexists in equilibrium in all three states of matter, i.e. ice, water and vapour is called triple point of water.

Triple point of water is 273 K temperature and 0.46 cm of mercury pressure.

Solar Constant

The amount of heat received from the sun by one square centimetre area of a surface placed normally to the sun rays at mean distance of the earth from the sun is known as solar constant. It is denoted by S.

$$S = \left(\frac{r}{R}\right)^2 \sigma T^4$$

Here, r is the radius of sun and R is the mean distance of earth from the centre of sun. Value of solar constant is 1.937 cal cm^{-2} min^{-1}.

CHAPTER 15

Ray Optics

Two Laws of Reflection

(i) $\angle i = \angle r$

(ii) Incident ray, reflected ray and normal lie on same plane.

Note *(i) While applying $\angle i = \angle r$, normal at point of incidence is very important. In case of shperical surface, normal at every point passes through the centre of curvature.*

(ii) If three unit vectors are given in the direction of incident ray, reflected ray and normal, then these three vectors should be coplanar or their scalar triple product should be zero.

Reflection from Plane Surface (or Plane Mirror)

- $v = -u$

  Here, v and u are measured from the plane surface. Two conclusions can be drawn from this equation.

  (i) Negative sign implies that object and image are on opposite sides of the mirror. So, if object is real, then image is virtual and *vice-versa*.

  (ii) $|v| = |u|$

  This implies that, perpendicular distance of the object from the mirror is equal to the perpendicular distance of image from the mirror.

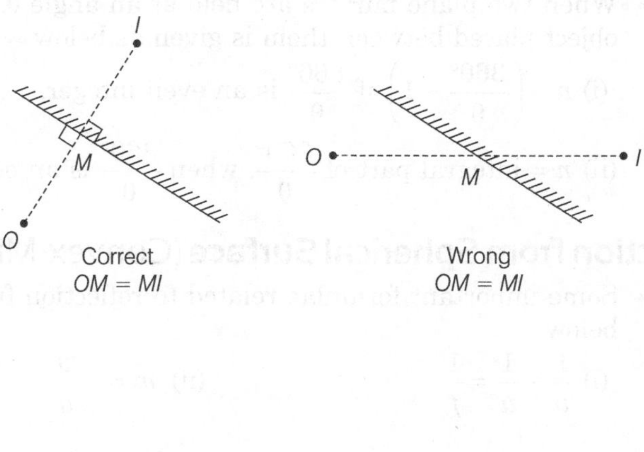

- **Ray diagram**

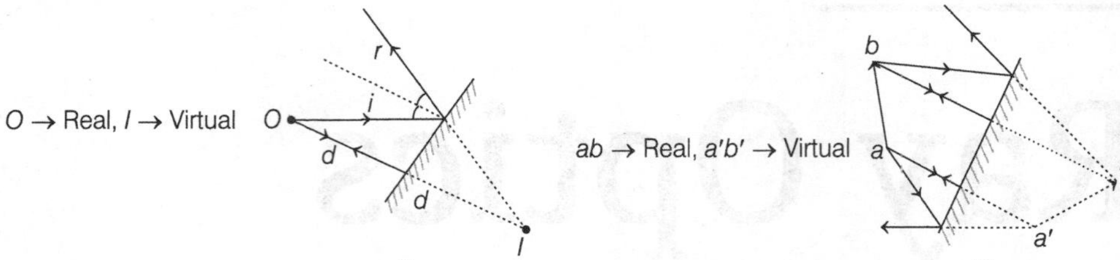

(i) (ii)

- **Field of view**

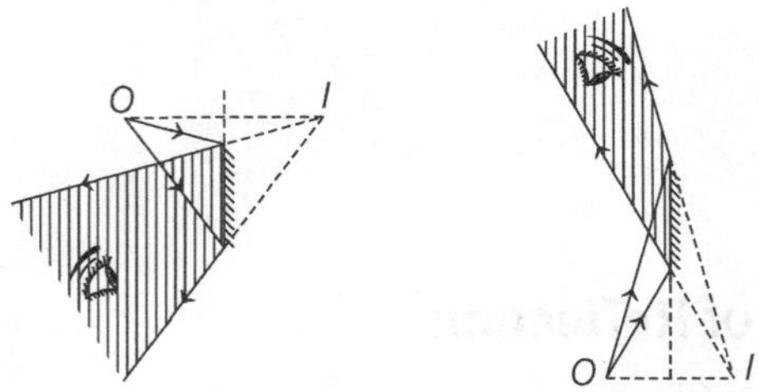

- Suppose a mirror is rotated by an angle θ (say anti-clockwise), keeping the incident ray fixed, then the reflected ray rotates by 2θ along the same direction, i.e. anti-clockwise.
- **Object and image velocity**
  (i) Image speed is equal to the object speed.
  (ii) Image velocity and object velocity make same angle from the plane mirror on two opposite sides of the mirror.
  (iii) Components of velocities which are along the mirror are equal.
  (iv) Components of velocities which are perpendicular to the mirror are equal and opposite.

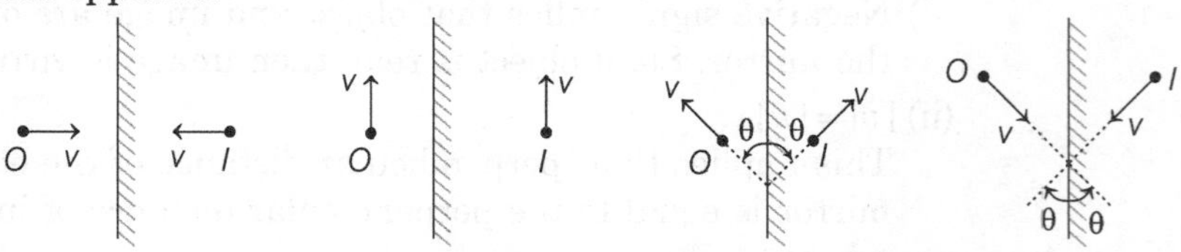

- When two plane mirrors are held at an angle θ, the number of images of an object placed between them is given as below
  (i) $n = \left(\frac{360°}{\theta} - 1\right)$, if $\frac{360°}{\theta}$ is an even integer.
  (ii) n = integral part of $\frac{360°}{\theta}$, when $\frac{360°}{\theta}$ is an odd integer.

Reflection from Spherical Surface (Convex Mirror or Concave Mirror)

- Some important formulae related to reflection from spherical surface are given below
  (i) $\frac{1}{v} + \frac{1}{u} = \frac{1}{f}$ (ii) $m = -\frac{v}{u}$

(iii) $f = \dfrac{R}{2}$ (iv) $P(\text{dioptre}) = \dfrac{-1}{f(\text{m})}$

- In case of spherical mirrors, if object distance x_1 and image distance x_2 are measured from focus instead of pole, then $u = (f + x_1)$ and $v = (f + x_2)$ and the mirror formula

$$\frac{1}{v} + \frac{1}{u} = \frac{1}{f} \text{ reduces to, } \frac{1}{f + x_2} + \frac{1}{f + x_1} = \frac{1}{f}$$

which on simplification gives, $x_1 x_2 = f^2$

This formula is called **Newton's formula.**

This formula applies to a lens also, but in that case x_1 is the object distance from first focus and x_2 is the image distance from second focus.

- **Image velocity**

***Case* 1** Along the principal axis, $\mathbf{v}_I = -m^2 \mathbf{v}_O$.

Here, negative sign implies that object and image always travel in opposite directions.

***Case* 2** Perpendicular to principal axis, $\mathbf{v}_I = m\, \mathbf{v}_O$.

Here, m has to be substituted with sign. If m is positive, then $\mathbf{v}_I$ and $\mathbf{v}_O$ travel in same direction.

If m is negative, then they travel in opposite directions.

- **Images formed by spherical mirrors**

***Case* 1 Concave mirror**

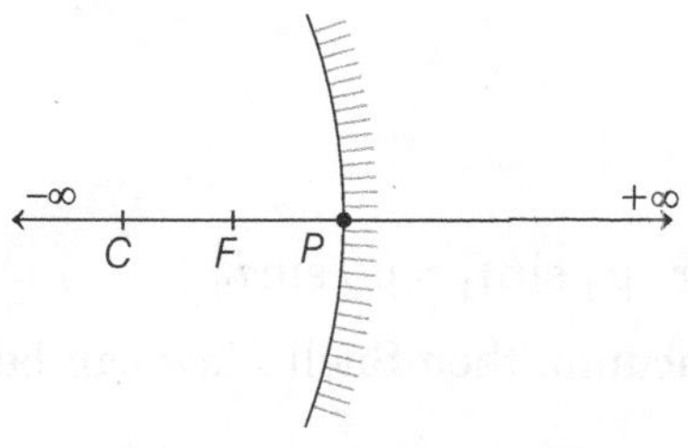

Object position	Image position	Nature of Image	Object and image speed
P	P	–	–
F	$\pm\infty$	–	–
C	C	Real, inverted and same size	$v_I = v_o$
Between P and F	Between P and $+\infty$	Virtual, erect and magnified	$v_I > v_o$
Between F and C	Between $-\infty$ and C	Real, inverted and magnified	$v_I > v_o$
Between C and $-\infty$	Between C and F	Real, inverted and diminished	$v_o > v_I$

Note *(i) The above table is only for real objects lying in front of the mirror for which u is negative.*
(ii) v_I *and* v_O *are image and object speeds.*
(iii) From the above table we can see that image and object always travel in opposite directions as long as they are moving along the principal axis.

***Case* 2 Convex mirror** For real objects there is only one case.

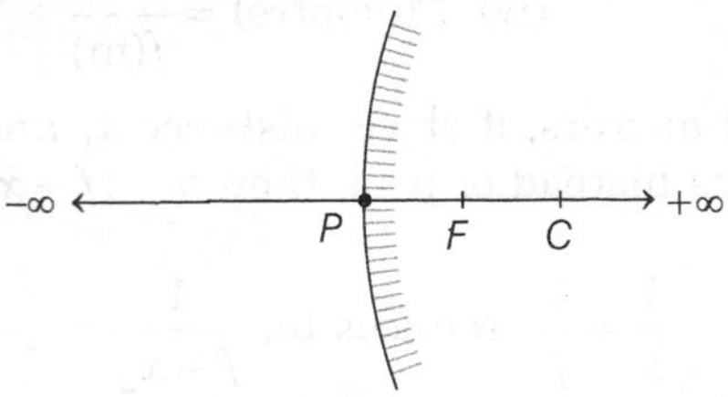

If object lies between P and $-\infty$, then image lies between P and F. Image is virtual, erect and diminished.

Object speed is greater than the image speed and they travel in opposite directions (along the principal axis).

- Following table represents the nature of the image and magnification for the object at different positions for concave and convex mirrors.

Value of m	Nature of image	Type of mirror	Object position
-4	Inverted, real and magnified	Concave	Between F and C
-1	Inverted, real and same size	Concave	At C
$-\frac{1}{2}$	Inverted, real and diminished	Concave	Between C and $-\infty$
$+3$	Erect, virtual and magnified	Concave	Between P and F
$+1$	Erect, virtual and same size	Plane	For all positions
$+\frac{1}{2}$	Erect, virtual and diminished	Convex	Between P and $-\infty$

Refraction of Light

- Snell's law is
  $\mu \sin i = \text{constant}$ or $\mu_1 \sin i_1 = \mu_2 \sin i_2$
- If one medium is vacuum, then Snell's law can be written as, $\mu = \dfrac{\sin i}{\sin r}$

  Here, i is angle of ray of light with normal in vacuum (or in air) and r the angle in medium.
- μ = absolute refractive index of a medium $= \dfrac{c}{v}$

  Here, c = speed of light in vacuum
  and v = speed of light in the medium.
- ${}_1\mu_2$ = Refractive index of 2 with respect to 1.
  $$= \frac{\mu_2}{\mu_1} = \frac{v_1}{v_2} = \frac{\lambda_1}{\lambda_2} = \frac{\sin i_1}{\sin i_2}$$
- ${}_1\mu_2 = \dfrac{1}{{}_2\mu_1}$
- ${}_1\mu_2 \times {}_2\mu_3 \times {}_3\mu_1 = 1$

- In $\mu = \dfrac{\sin i}{\sin r}$, angle i is not always the angle of incidence but it is the angle of ray of light in air (with normal).

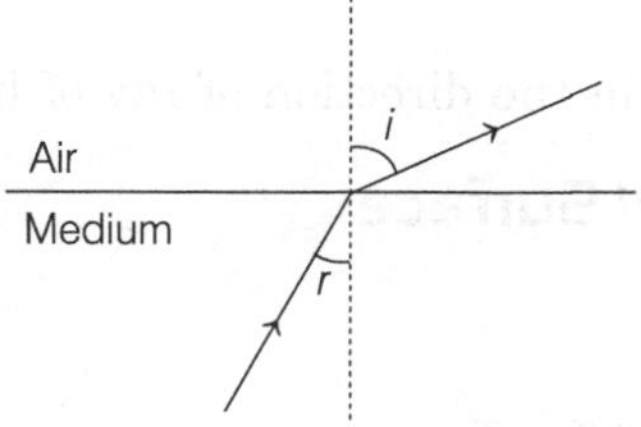

In the figure, ray of light is travelling from medium to air. So, angle of incidence is actually r. But we have to take i angle in air and now we can apply $\mu = \dfrac{\sin i}{\sin r}$.

- In $\mu = \dfrac{\sin i}{\sin r}$, if i is changed, then r angle also changes. But $\dfrac{\sin i}{\sin r}$ remains constant and this constant is called refractive index of that medium.
- $\mu = \dfrac{\sin i}{\sin r}$ can be applied for any pair of angles i and r except the normal incidence for which $\angle i = \angle r = 0°$ and $\mu = \dfrac{\sin i}{\sin r}$ is an indeterminant form.
- $\mu = \mu_0 + \dfrac{A}{\lambda^2}$ (Cauchy's formula)

Note *The colour of light depends on its frequency, not its wavelength. Therefore, in refraction colour will not change.*

Single Refraction from a Plane Surface

- $v = \left(\dfrac{\mu_2}{\mu_1}\right) \cdot u$
- Object and image are on same side. If object is real, then image is virtual and *vice-versa*.
- If object is lying in a denser medium and you observe it from a rarer medium, then its image appears nearer. If object is lying in a rarer medium and you observe it from a denser medium, then it appears farther. If rarer medium is vacuum, then this increase or decrease in distance is μ times. Hence,

$$d_{app} = \frac{d}{\mu} \text{ and } h_{app} = \mu h$$

- For more than one liquids, $d_{app} = \dfrac{d_1}{\mu_1} + \dfrac{d_2}{\mu_2}$
- Image speed, $v_I = \dfrac{\mu_2}{\mu_1}$ (object speed). But they travel in same direction.

Shift from a Glass Slab

- This is double refraction from two plane surfaces.
- Shift, $S = \left(1 - \frac{1}{\mu}\right) t$. Here, t is thickness of glass slab.
- This shift takes place in the direction of ray of light.

Refraction from Spherical Surface

- $\frac{\mu_2}{v} - \frac{\mu_1}{u} = \frac{\mu_2 - \mu_1}{R}$
- For plane surface, $R = \infty$.

$$\therefore \quad \frac{\mu_2}{v} - \frac{\mu_1}{u} = 0$$

or

$$v = \frac{\mu_2}{\mu_1} \cdot u$$

Lens Theory

- Any lens is made up of two surfaces, of which at least one surface should be spherical.
- All lens formulae have been derived for following two conditions
  (i) Lens thickness should be negligible.
  (ii) On both sides of the lens medium should be same.
- If neither of the above two conditions is satisfied, then we should apply refraction formula, $\frac{\mu_2}{v} - \frac{\mu_1}{u} = \frac{\mu_2 - \mu_1}{R}$ two times.
- If above two conditions are satisfied, then object can be placed on any side of the lens. Image distance comes out to be same for same object distance.
- $\frac{1}{f} = \left(\frac{\mu_2}{\mu_1} - 1\right)\left(\frac{1}{R_1} - \frac{1}{R_2}\right)$

  Here, μ_2 = refractive index of lens and μ_1 = refractive index of medium
- $\frac{1}{v} - \frac{1}{u} = \frac{1}{f}$
- Magnification of lens, $m = \frac{v}{u}$
- Power of lens, $P(\text{dioptre}) = \frac{+1}{f(\text{m})}$
- When two or more than two thin lenses in contact,

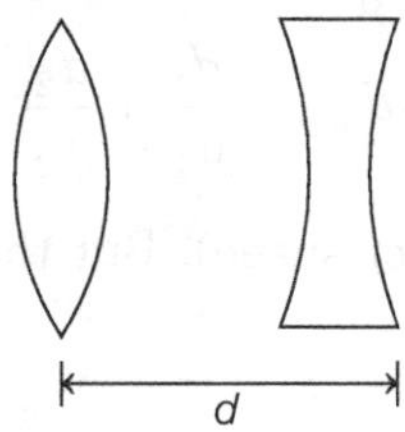

$$\frac{1}{F} = \frac{1}{f_1} + \frac{1}{f_2} \text{ or } P = P_1 + P_2$$

When two or more than two thin lenses at some distance,

$$\frac{1}{F} = \frac{1}{f_1} + \frac{1}{f_2} - \frac{d}{f_1 f_2} \quad \text{or} \quad P = P_1 + P_2 - dP_1P_2$$

- A silvered lens behaves as a mirror of focal length given by

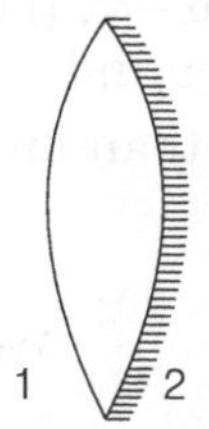

$$\frac{1}{f} = \frac{2(\mu_2/\mu_1)}{R_2} - \frac{2(\mu_2/\mu_1 - 1)}{R_1}$$

After finding focal length from this formula, we will apply mirror formula for finding the image position.

- **Image Formed by Lens**

***Case* 1 Convex lens**

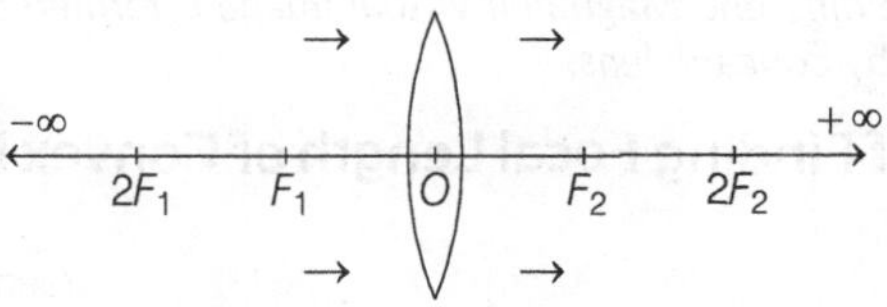

Object position	Image position	Nature of image	Object and image speed
At F_1	$\pm\ \infty$	–	–
At $2F_1$	At $2F_2$	Real, Inverted and same size	$v_I = v_o$
At $-\infty$	At F_2	–	–
Between O and F_1	Between O and $-\infty$	Virtual, Erect and magnified	$v_I > v_o$
Between F_1 and $2F_1$	Between $+\infty$ and $2F_2$	Real, Inverted and magnified	$v_I > v_o$
Between $2F_1$ and $-\infty$	Between $2F_2$ and F_2	Real, Inverted and diminished	$v_o > v_I$

Note (i) *The above table has been made only for real objects (lying between O and $-\infty$), object distance u for them is negative.*

(ii) *Since $|f_1| = |f_2|$ (when the conditions discussed above is satisfied). Therefore, F_1 and F_2 are sometimes denoted by F and $2F_1$ (or $2F_2$) by 2F.*

(iii) *If object is travelling along the principal axis, then image also travels along the principal axis in the same direction.*

***Case* 2 Concave lens** In case of concave lens, there is only one case for real objects.

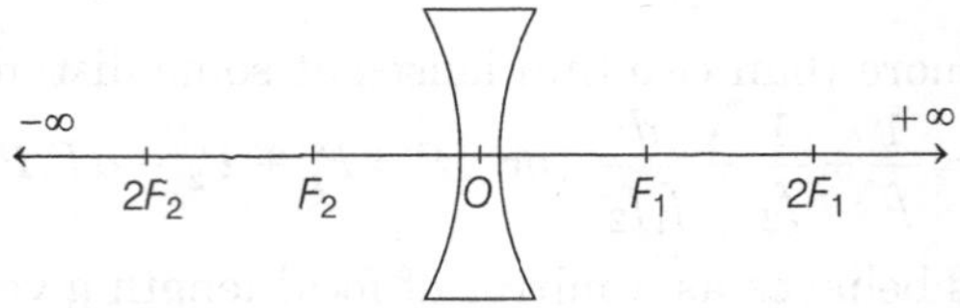

When object lies between O and $-\infty$, then image lies between O and F_2. Nature of image is virtual, erect and diminished.

Object speed is always greater than image speed ($v_o > v_I$). Both travel in the same direction.

Value of m	Nature of image	Type of lens	Object position
-3	Inverted, real and magnified	convex	Between F_1 and $2F_1$
-1	Inverted, real and same size	convex	at $2F_1$
$-\frac{1}{2}$	Inverted, real and diminished	convex	Between $2F_1$ and $-\infty$
$+2$	Erect, virtual and magnified	convex	Between O and F_1
$+\frac{1}{4}$	Erect, virtual and diminished	concave	Between O and $-\infty$

Note *For real objects, real image is formed only by convex lens. But virtual image is formed by both types of lenses. Their sizes are different. Magnified virtual image is formed by convex lens. Diminished virtual image is formed by concave lens.*

Displacement Method of Finding Focal Length of Convex Lens

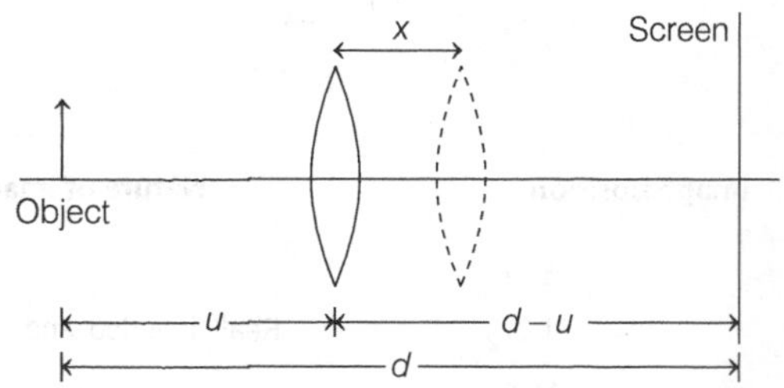

Object and screen are fixed. If $d > 4f$, there are two positions of lens between object and screen for which real image is formed on the screen.

Suppose I_1 is the image length in one position of the lens and I_2 the image length in second position, then object length O is given by

$$O = \sqrt{I_1 I_2}$$

Focal length of the lens is given by $f = \dfrac{d^2 - x^2}{4d}$

If one image is $\frac{1}{4}$th in size than other image will be 4 times in size.

or $$m_1 m_2 = 1$$

Total Internal Reflection (TIR)

- When a ray of light travels from a denser to a rarer medium with angle of incidence (i) > critical angle (θ_C), then no refraction takes place.
  Ray of light is 100% reflected. This phenomenon is called total internal reflection.
- Critical angle, $\theta_C = \sin^{-1}\left(\frac{\mu_R}{\mu_D}\right)$
  If rarer medium is vacuum (or air), then
  $$\theta_C = \sin^{-1}\left(\frac{1}{\mu}\right)$$
- More the value of refractive index μ, smaller is the value of critical angle θ_C and more are the chances of TIR.
- Optical fibres are also based on the phenomenon of total internal reflection. Optical fibres consist of several thousands of very long fine quality fibres of glass or quartz.
- The diameter of each fibre is of the order of 10^{-4} cm with refractive index of material being of the order of 1.5.
- Optical fibres are used in transmission and reception of electrical signals by converting them first into light signals.

Prism Theory

- $\angle A$ is called angle of prism or refracting angle of prism.

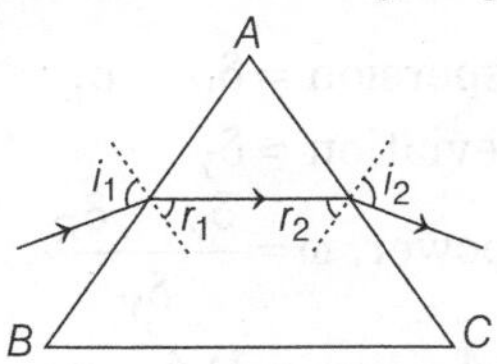

- $r_1 + r_2 = A$
- Deviation by prism, $\delta = (i_1 + i_2) - A \approx (\mu - 1)A$, if A and i_1 are small.
- At minimum deviation,

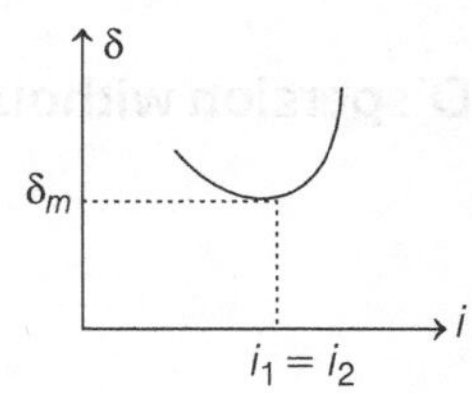

$$r_1 = r_2 = \frac{A}{2} \text{ and } i_1 = i_2 = \left(\frac{A + \delta_m}{2}\right)$$

- $\mu = \dfrac{\sin\left(\dfrac{A + \delta_m}{2}\right)}{\sin\left(\dfrac{A}{2}\right)}$
- If $A > 2\theta_C$, ray of light will not emerge from face AC. It gets total internally reflected there.
- **Deviation (δ)**
  (i) In reflection $\delta = 180 - 2i = 180 - 2r$
  (ii) In refraction $\delta = i \sim r$

Dispersion of Light by a Prism

- **Cauchy's formula**

$$\mu_\lambda = \mu_0 + \frac{A}{\lambda^2} + \frac{B}{\lambda^4} + \dots$$

$$\approx \mu_0 + \frac{A}{\lambda^2}$$

and $\mu_V > \mu_R$

- This spreading of light into its colour components is called dispersion of light.

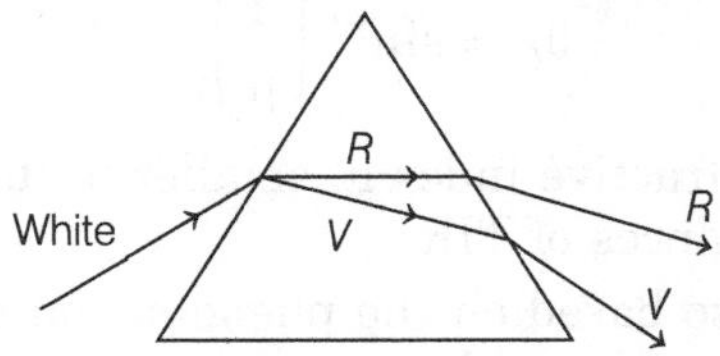

- **Dispersive power**

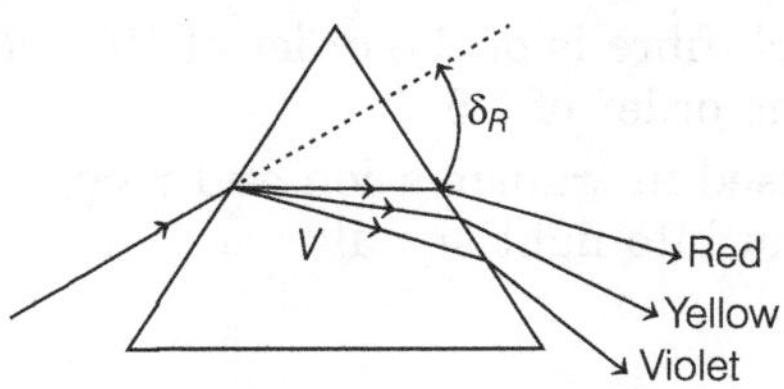

$$\text{Angular dispersion} = \delta_V - \delta_R$$

$$\text{Average deviation} = \delta_Y$$

$$\text{Dispersive power, } \omega = \frac{\delta_V - \delta_R}{\delta_Y}$$

or
$$\omega = \frac{(\mu_V - 1)A - (\mu_R - 1)A}{(\mu_Y - 1)A} = \frac{\mu_V - \mu_R}{(\mu_Y - 1)}$$

$$\omega_{\text{CROWN}} < \omega_{\text{FLINT}}$$

Dispersion without Deviation and Deviation without Dispersion

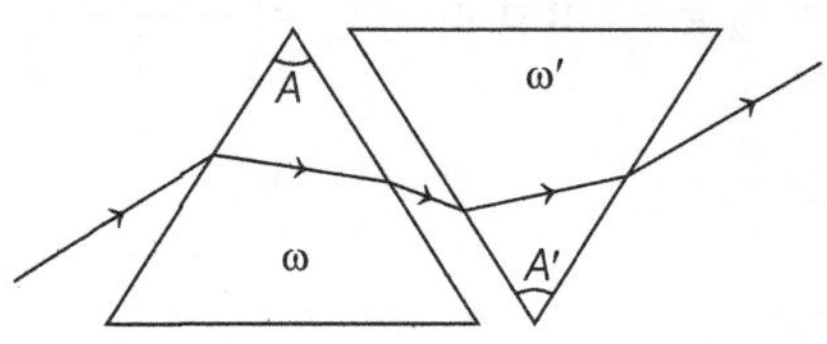

- Average deviation,

$$\delta_Y = (\mu_Y - 1)A - (\mu'_Y - 1)A' \quad \text{...(i)}$$

- Net angular dispersion,

$$\delta_V - \delta_R = (\mu_V - \mu_R)A - (\mu'_V - \mu'_R)A$$

$$\delta_V - \delta_R = \omega(\mu_Y - 1)A - \omega'(\mu'_Y - 1)A' \quad \text{...(ii)}$$

- Dispersion without deviation

$$\delta_Y = 0 \quad \Rightarrow \quad \frac{A}{A'} = \left(\frac{\mu_Y' - 1}{\mu_Y - 1}\right)$$

Substituting in Eq. (ii), we get

Net angular dispersion $= (\delta_V - \delta_R) = (\mu_Y - 1)\, A(\omega - \omega')$

- Deviation without dispersion, $\delta_V - \lambda_R = 0$

$$\Rightarrow \qquad \frac{A,}{A} = \frac{(\mu_Y' - 1)\,\omega'}{(\mu_Y - 1)\,\omega} = \left(\frac{\mu_V' - \mu_R'}{\mu_V - \mu_R}\right)$$

Substituting in Eq. (i), we have net deviation,

$$\delta_Y = (\mu_Y - 1)A\left(1 - \frac{\omega}{\omega'}\right)$$

Optical Instruments

Name of optical instruments	M	L	M_∞	M_D	L_∞	L_D
Simple microscope	$\frac{D}{u_o}$	—	$\frac{D}{f}$	$1 + \frac{D}{f}$	—	—
Compound microscope	$\frac{v_o}{u_o}\frac{D}{u_e}$	$v_o + u_e$	$\frac{v_o}{u_o}\frac{D}{f_e}$	$\frac{v_o}{u_o}\left(1 + \frac{D}{f_e}\right)$	$v_o + f_e$	$v_o + \frac{Df_e}{D + f_e}$
Astronomical telescope	$\frac{f_o}{u_e}$	$f_o + u_e$	$\frac{f_o}{f_e}$	$\frac{f_o}{f_e}\left(1 + \frac{f_e}{D}\right)$	$f_o + f_e$	$f_o + \frac{Df_e}{D + f_e}$
Terrestrial telescope	— do —	$f_o + 4f + u_e$	— do —	— do —	$f_o + 4f + f_e$	$f_o + 4f + \frac{Df_e}{D + f_e}$
Galilean telescope	$\frac{f_o}{u_e}$	$f_o - u_e$	$\frac{f_o}{f_e}$	$\frac{f_o}{f_e}\left(1 - \frac{f_e}{D}\right)$	$f_o - f_e$	$f_o - \frac{f_eD}{D - f_e}$

Note *(i) In all above formulae of M, we are considering only the magnitude of M.*

(ii) For telescopes, formulae have been derived when the object is at infinity. For the object at some finite distance different formulae will have to be derived.

Resolving Power of Microscope and Telescope

Microscope

- **Limit of resolution (say Δd)** Minimum distance between two point objects which can be seen separately is called limit of resolution.
- If we see an object with the help of a microscope, then obviously Δd will be less as compared to naked eye and its resolving power will be more.

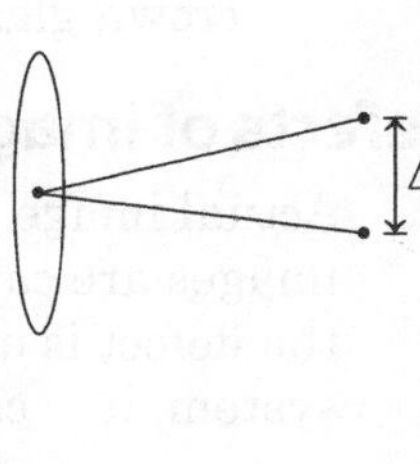

Thus, resolving power, $R = \frac{1}{\Delta d} = \frac{2\mu \sin\theta}{\lambda}$

Here, μ = refractive index of medium

In oil, μ will increase, λ will decrease.

So, R will increase.

Telescope

$\Delta\theta$ = minimum angular separation between two objects which can be seen (resolved) by a telescope.

$$R = \frac{1}{\Delta\theta} = \frac{a}{1.22\lambda}$$

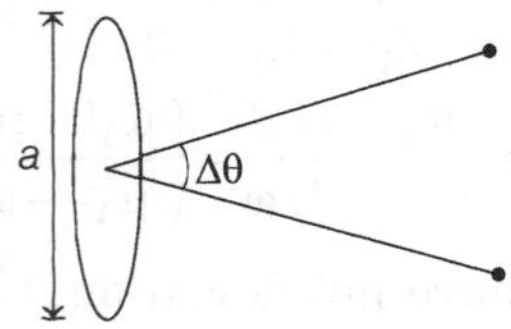

Condition of Achromatism

To get achromatism, we use a pair of two lenses in contact. For two thin lenses in contact, we have

$$\therefore \quad \frac{\omega_1}{f_1} + \frac{\omega_2}{f_2} = 0$$

This is the **condition** of **achromatism.** From the condition of achromatism, following conclusions can be drawn

(i) As ω_1 and ω_2 are positive quantities, f_1 and f_2 should have opposite signs, i.e. if one lens is convex, the other must be concave.

(ii) $\omega_1 = \omega_2$, means both the lenses are of same material. Then,

$$\frac{1}{f_1} + \frac{1}{f_2} = 0$$

or
$$\frac{1}{F} = 0$$

or
$$F = \infty$$

Thus, the combination behaves as a plane glass plate. So, we can conclude that both the lenses should be of different materials or $\omega_1 \neq \omega_2$

(iii) Dispersive power of crown glass (ω_C) is less than that of flint glass (ω_F).

(iv) If we want the combination to behave as a convergent lens, then convex lens should have lesser focal length or its dispersive power should be more. Thus, convex lens should be made of flint glass and concave lens of crown. Similarly, for the combination to behave as diverging lens, convex is made of crown glass and concave of flint glass.

Defects of Images

Actual image formed by an optical system is usually imperfect. The defects of images are called aberrations. The defect may be due to light or optical system. If the defect is due to light it is called chromatic aberration and if due to optical system, it is called monochromatic aberration.

(i) **Chromatic aberration** The image of an object formed by a lens is usually coloured and blurred. This defect of image is called chromatic aberration. This defect arises due to the fact that focal length of a lens is different for different colours.

For a lens, $\frac{1}{f} = (\mu - 1)\left(\frac{1}{R_1} - \frac{1}{R_2}\right)$

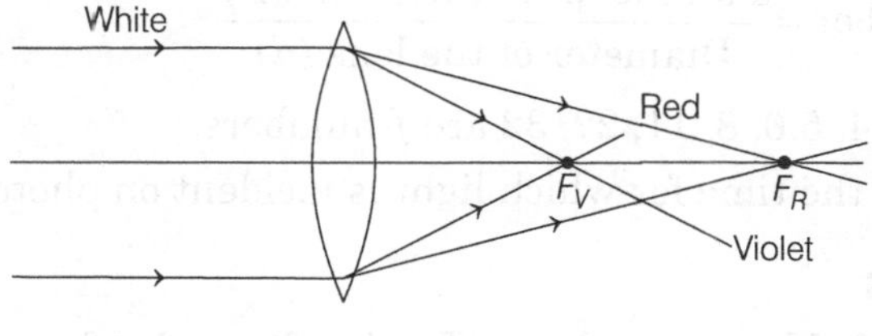

As μ is maximum for violet while minimum for red, violet is focused nearest to the lens while red farthest from it.

The difference between f_R and f_V is a measure of longitudinal chromatic aberration. Thus,

$$\text{LCA} = f_R - f_V$$

(ii) **Monochromatic aberration** This is the defect in image due to optical system. Monochromatic aberration is of many types such as, spherical, coma, distortion, curvature and astigmatism. Here, we shall limit ourselves to spherical aberration only.

Spherical aberration arises due to spherical nature of lens (or mirror).

The paraxial rays (close to optic axis) get focused at I_P and marginal rays (away from the optic axis) are focused at I_M. Thus, image of a point object O is not a point.

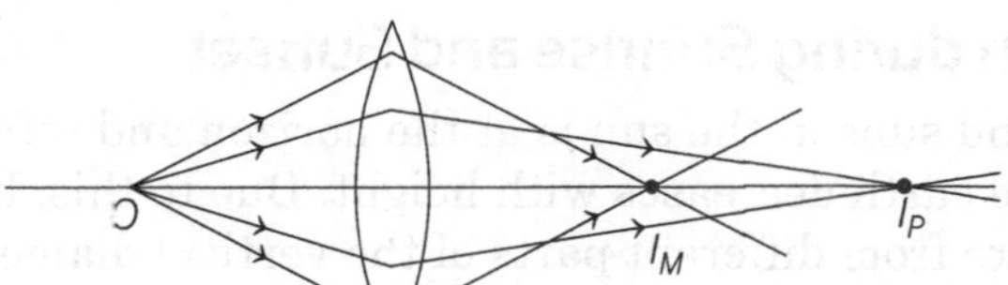

The inability of the lens to form a point image of an axial point object is called spherical aberration.

The Eye

- Focal length of eye lens is about 2.5 cm.
- Power of accommodation for normal eye $D = 25$ cm.
- In myopia or short-sightedness, concave lens is used.
- In hypermetropia or far-sightedness, convex lens is used.
- In astigmatism, cylindrical lens is used.

Camera

A photograph camera consists of a light proof box, at one end of which a converging lens system is fitted. A light sensitive film is fixed at the other end of the box, opposite to the lens system. A real inverted image of the object is formed on the film by the lens system.

f-Number for a camera The f-number represents the size of the aperture.

$$f\text{-number} = \frac{\text{Focal length of the lens } (F)}{\text{Diameter of the lens } (d)}$$

Generally 2, 2.8, 4, 5.6, 8, 11, 22, 32 are f-numbers.

Exposure time is the time for which light is incident on photographic film.

Twinkling of Stars

Earth is surrounded by atmosphere. The density and refractive index both goes on increasing on moving from upper level of atmosphere towards the earth. All the stars that we see are very far away from us and therefore they can be considered as point like sources of light.

So, when the light from the stars enter the atmosphere they pass through denser and rarer mediums, hence undergoing refraction continuously. As a result of this refraction, the apparent positions of the stars are changing continuously and it seems that the stars are twinkling.

This is one of the reasons the hubble telescope is so successful in space, there is no atmosphere to make the stars twinkle, allowing it much better image to be obtained.

Note *Planets are much closer and appear as broader disks and so the refraction from one point on the planet gets cancelled, on the average, from another part of the disk, hence the fluctuations get smoothened and we don't see them twinkling.*

Oval Shape of Sun during Sunrise and Sunset

During sunrise and sunset, the sun is at the horizon and refractive index of the atmosphere of the earth decreases with height. Due to this, light reaching the earth's atmosphere from different parts of the vertical diameter of the sun enters at different heights in earth's atmosphere and so bend unequally.

Due to this unequal bending of light from the vertical diameter, the image of the sun gets distorted and it appears oval and larger. However, at noon when the sun is overhead, then due to normal incidence of light there is no bending of light and hence, the sun appears circular.

Advanced Sunrise and Delayed Sunset

It's all about refraction. This is because when light enters from vacuum to earth's atmosphere, it basically enters from rarer to denser medium and bends towards horizon. During sunrise, when the sun is just below the horizon, our atmosphere causes the light rays to bend and we see the sun early.

Similarly, at sunset, the apparent position of the sun is visible to us and not the actual position due to the same bending of light ray's effect.

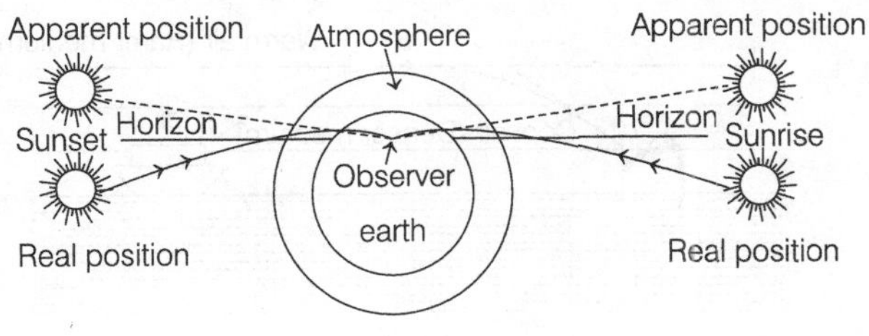

Mirage and Looming Mirage

Mirage is an optical illusion which occurs usually in deserts on hot summer days due to atmospheric refraction and total internal reflection of light rays.

In mirage, the object such as a tree appears to be inverted as if it is situated on a bank of a pond of water.

Actually on a hot summer day, the layers of air near the surface of earth become very hot and hence behave as optically rarer medium. On the other hand, the upper layers of air are comparatively cool and hence behave as optically denser medium.

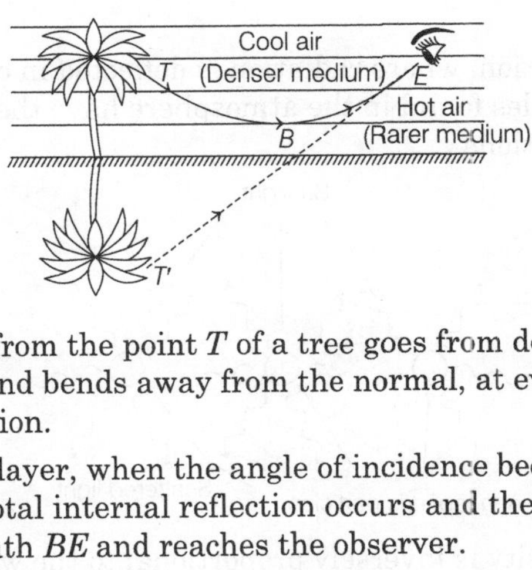

Now, a ray of light from the point T of a tree goes from denser to rarer medium along the path TB and bends away from the normal, at every layer due to atmospheric refraction.

But at a particular layer, when the angle of incidence becomes greater than the critical angle, the total internal reflection occurs and the totally reflected ray travels along the path BE and reaches the observer.

Since, we can see the light rays only in straight line path, so the reflected ray BE appears to be coming from the point T' to the observer. Due to this, an inverted image of the tree is formed below its actual position. And this inverted image of the tree creates the impression as if the reflection of tree is taking place from a pond full of water.

Looming

Looming is also an optical illusion which occurs usually in very cold regions. In looming, a distant object such as a ship moving in polar areas appears to be hanging in mid air due to atmospheric refraction and the total internal reflection of light rays.

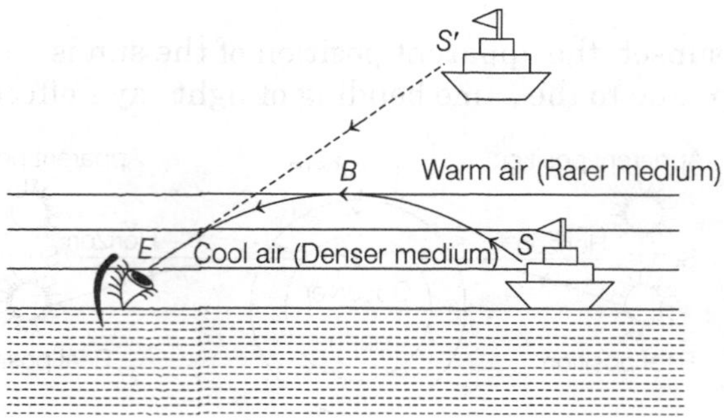

In polar regions, the layers of air near the surface of earth are very cold and hence behave as optically denser medium, whereas the upper layers of air are comparatively warm and hence behave as optically rarer medium.

Now, a ray of light coming from point S of ship goes from denser to rarer medium along the path SB and bends away from the normal, at every layer due to atmospheric refraction.

But, at a particular layer, when the angle of incidence becomes greater than the critical angle, the total internal reflection occurs and the reflected ray BE appears to be coming from the point S' to the observer. Due to this, the observer sees a virtual image of the ship at position S'.

Scattering of Light

Different from reflection, where radiation is deflected in one direction, some particles and molecules found in the atmosphere have the ability to scatter solar radiation in all directions.

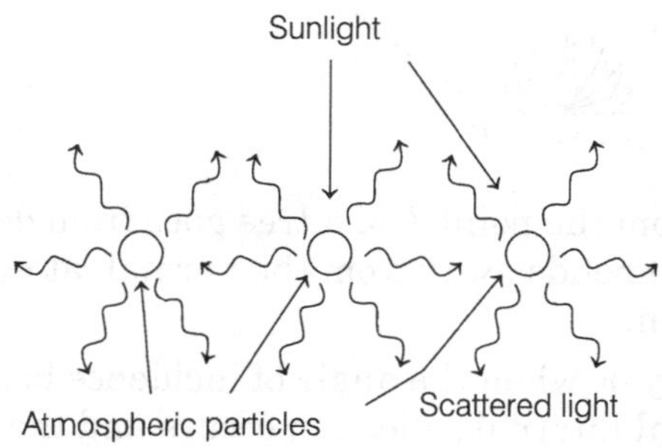

The scattering intensity is inversely proportional to the wavelength to the fourth power $\left(I \propto \frac{1}{\lambda^4}\right)$. This phenomenon is known as Rayleigh scattering.

Air molecules, like oxygen and nitrogen, are small in size and are more effective at scattering shorter wavelengths of light (blue and violet).

Due to the nitrogen-oxygen composition of the atmosphere, blue light is scattered 14.4 times as much as red light. The scattering causes a clear sky to look blue.

Sunsets are red because the sunlight has to penetrate a greater thickness of atmosphere and so most of the blue light is removed by scattering, which leave red light predominantly.

CHAPTER 16

Wave Optics

According to wave theory of light, the light is a form of energy which travels through a medium in the form of transverse wave. The speed of light in a medium depends upon the nature of medium.

Newton's Corpuscular Theory

- Light consists of very small invisible elastic particles which travel in vacuum with a speed of 3×10^8 m/s.
- The theory could explain reflection and refraction.
- The size of corpuscular of different colours of light are different.
- It could not explain interference, diffraction, polarisation, photoelectric effect and Compton effect. The theory failed as it could not explain why light travels faster in a rarer medium than in a denser medium.

Wavefront

- An imaginary surface on which all particles are in same phase. It is usually perpendicular to the direction of propagation of light.
- Some of the wavefronts are shown by dotted lines in the figures given below.

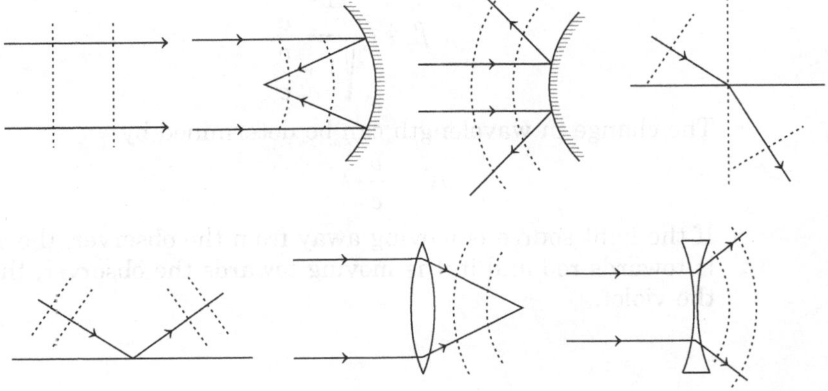

Huygens' Wave Theory

- Light travels in a medium in the form of wavefront.
- A wavefront is the locus of all the particles vibrating in same phase.

  All particles on a wavefront behave as a secondary source of light, which emits secondary wavelets.
- The envelope of secondary wavelets represents the new position of a wavefront.

Huygens' Principle

- Every point on given wavefront (called primary wavefront) acts as a fresh source of new disturbance called secondary wavelets.
- The secondary wavelets travels in all the directions with the speed of light in the medium.
- A surface touching these secondary wavelets tangentially in the forward direction at any instant gives the new (secondary) wavefront of that instant.
- Refraction and reflection laws can be explained on the basis of Huygens' principle.

Doppler Effect in Light

- Light waves also show Doppler's effect. If a light source is moving away from a stationary observer, then the frequency of light waves appear to be decreased and wavelength appear to be increased and *vice-versa.*
- If the light source or the observer is moving with a velocity v such that the distance between them is decreasing, then the apparent frequency of the source will be given by

$$f' = f\sqrt{\frac{1+\frac{v}{c}}{1-\frac{v}{c}}}$$

- If the distance between the light source and the observer is increasing, then the apparent frequency of the source is given by

$$f' = f\sqrt{\frac{1-\frac{v}{c}}{1+\frac{v}{c}}}$$

- The change in wavelength can be determined by

$$\Delta\lambda = \frac{v}{c}\cdot\lambda$$

- If the light source is moving away from the observer, the shift in the spectrum is towards red and if it is moving towards the observer, the shift is towards the violet.

Interference

Different Expressions of Path Difference and Their Limitations

- If $d << D$, then $\Delta x = S_2P - S_1P = d \sin\theta$

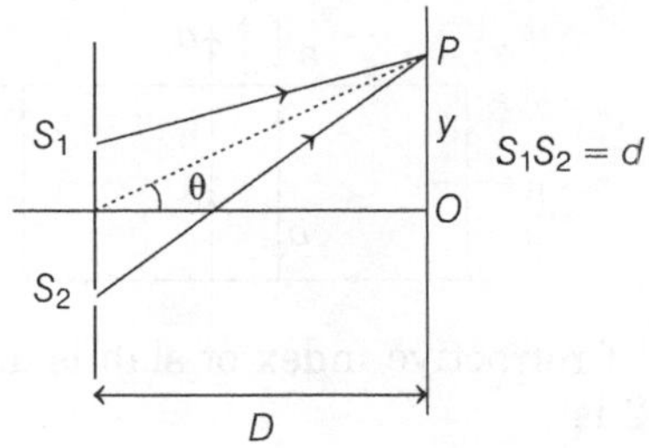

- In the above figure, if $d << D$ and θ is very small, then $\Delta x = S_2P - S_1P = \dfrac{yd}{D}$
- If $d << D$, then $\Delta x = S_1P - S_2P = d\cos\theta$

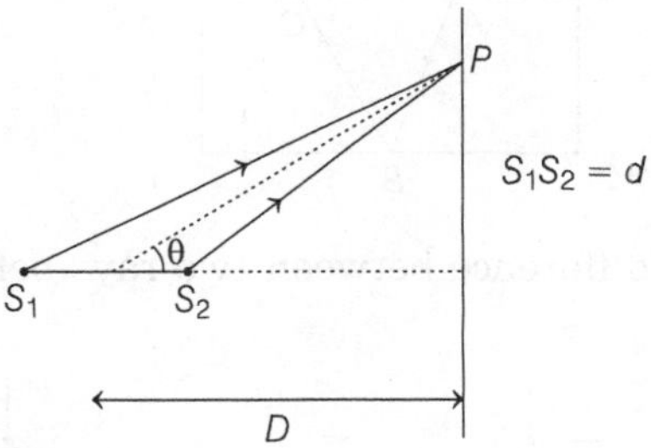

- If refractive index of slab is μ and thickness is t, then after emerging from the slab, path difference between rays 1 and 2 is,

$$\Delta x = (\mu - 1)\, t$$

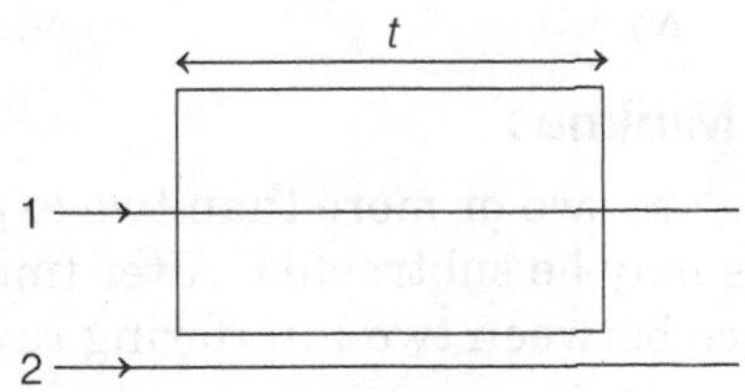

Note *If refractive index of slab is μ_2 and the medium in which this slab is kept is μ_1, then this path difference is given by*

$$\Delta x = \left(\frac{\mu_2}{\mu_1} - 1\right) t$$

- In the figure shown, a is the incident ray, e is reflected from top surface of the slab but d comes after reflecting from bottom surface. Then, the path difference between e and d is

$$\Delta x = 2\,\mu t$$

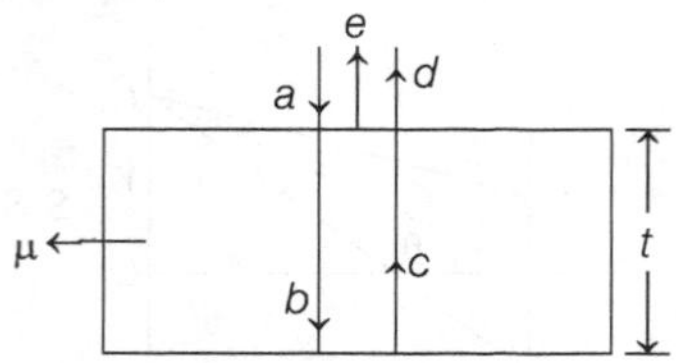

- In the figure shown, if refractive index of slab is μ, then path difference between rays 1 and 2 is

$$\Delta x = \mu(ABC) - AD$$

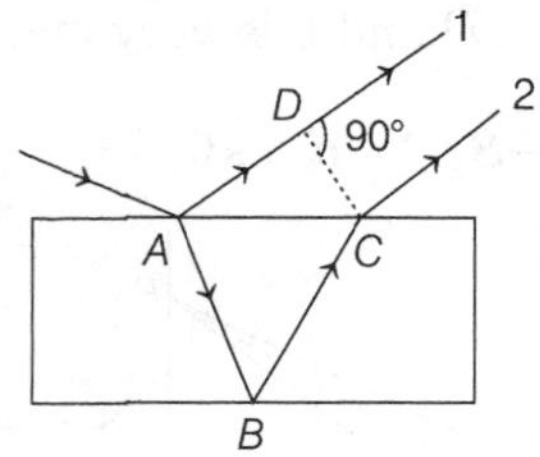

- Δx_i means the path difference between two rays before reaching the slits S_1 and S_2.

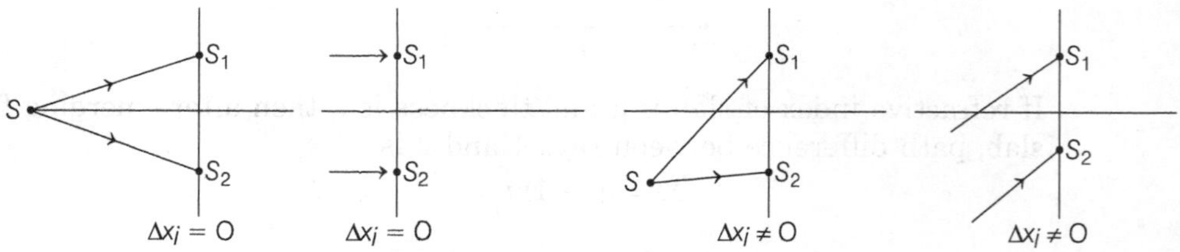

Conditions of Maximas and Minimas

In some problem, if there are two or more than two expressions of Δx, then some may be added and others may be subtracted. After finding the net path difference, check the phase difference between two interfering rays.

If $\Delta\phi = 0°$, then

$$\Delta x_{net} = n\lambda,\ n = 0, 1, 2, \dots \text{ for maximum intensity}$$

$$= (2n-1)\frac{\lambda}{2},\ n = 1, 2, \dots \text{ for minimum intensity}$$

If $\Delta\phi = \pi$, then

$$\Delta x_{net} = (2n-1)\frac{\lambda}{2} \text{ for maximum intensity}$$

and

$$= n\lambda \text{ for minimum intensity}$$

Young's Double Slit Experiment

- It is an experiment of interference in light.
- For interference sources must be coherent, interference is based on principle of superposition.
- y-coordinate of nth order maxima (or bright fringe) is

$$y_n = \frac{n\lambda D}{d} \qquad \text{(where, } n = 0, 1, 2, \ldots)$$

and y-coordinate of nth order minima (or dark fringe) is

$$y_n = (2n - 1)\frac{\lambda D}{2\,d} \qquad \text{(where, } n = 1, 2, 3, \ldots)$$

- Fringe shift due to a glass slab in the path of one of the slits

$$S = \frac{(\mu - 1)\,tD}{d}$$

This shift takes place on that side, where slab is placed.
This fringe shift is independent of n (the order of fringe) and λ (the wavelength of light).

- Number of fringes shifted, $N = \dfrac{\text{shift}}{\text{fringe width}} = \dfrac{(\mu - 1)\,t}{\lambda}$
- Fringe width $W = \dfrac{\lambda D}{d}$ i.e. $W \propto \lambda$

This implies that, if YDSE appratus is immersed in a liquid of refractive index μ, then wavelength and therefore, fringe width decreases μ times.
Further, if white light is used in YDSE, then coloured fringes are obtained as

$$\lambda_{red} > \lambda_{violet}$$

Hence, $\qquad W_{red} > W_{violet}$

But centre point is white because at that point, net path difference is zero for all wavelengths. Hence, all wavelengths interfere constructively.

- Angular fringe width $\theta = \dfrac{\lambda}{d}$. This is the angle subtended by one fringe at centre of two slits.
- Ratio of maximum intensity and minimum intensity in interference is given by

$$\frac{I_{max}}{I_{min}} = \left(\frac{\sqrt{I_1} + \sqrt{I_2}}{\sqrt{I_1} - \sqrt{I_2}}\right)^2 = \left(\frac{\sqrt{I_1/I_2} + 1}{\sqrt{I_1/I_2} - 1}\right)^2$$

$$= \left(\frac{A_1/A_2 + 1}{A_1/A_2 - 1}\right)^2 = \left(\frac{A_1 + A_2}{A_1 - A_2}\right)^2 = \left(\frac{A_{max}}{A_{min}}\right)^2$$

- If two individual intensities at any point P is assumed to be same (say I_0), then resultant intensity at P is given by

$$I = 4\,I_0 \cos^2\left(\frac{\phi}{2}\right)$$

where, ϕ is the phase difference between two interfering waves at P.

$$\phi = \frac{2\pi}{\lambda} \cdot \Delta x \quad \text{and} \quad \Delta x = d \sin \theta \quad \text{or} \quad \frac{yd}{D}$$

- In YDSE, if one slit is closed, then we will not get interference on screen and intensity at every point is almost uniform ($= I_0$) and this is due to only one slit.
- In YDSE, if both sources are incoherent, then again we will not get interference and intensity at every point is again uniform ($= I_1 + I_2$ or $2I_0$).
- In YDSE, if width of one slit is slightly increased, then I_{max} and I_{min} both will increase. This is because intensity due to the slit of increased width will increase

or $\qquad I_1 = I_0$ but $I_2 = nI_0 \qquad$ (where, $n > 1$)

$\therefore \qquad I_{min} = (\sqrt{I_1} - \sqrt{I_2})^2 > 0$

and $\qquad I_{max} = (\sqrt{I_1} + \sqrt{I_2})^2 > 4I_0$

- **Shape of fringes on complete screen** In the above discussion, we have observed the fringe pattern on the centre line of the screen.

  On this centre line, maximas (like y_0, y_1, y_2 etc.) or minimas (like y_1, y_2 etc.) are just points.

  But on the whole screen fringes make a curve and this curve is a locus of points, where path difference from two slits is a constant.

  or $\qquad \Delta x = S_1P - S_2P = c \qquad$ ... (i)

  $c = 0$, gives y_0 fringe. Similarly, $c = \pm \lambda$ gives y_1 or y_1' fringe etc.

  Now on the centre line, Δx was a function of only one variable coordinate $y \left(\Delta x = \frac{yd}{D} \right)$.

  But on the whole screen, it will become a function of two variable coordinates (say y and z). Therefore, Eq. (i) becomes $\Delta x = f(y, z) = c \qquad$ ... (ii)

  After proper calculations, we can show that this comes out to be a family of curves of hyperbolas. Thus, on the complete screen, fringes are of the shape of hyperbolas.

  For $c = 0$, this hyperbola converts into a straight line. Hence, only y_0 fringe is a straight line. The fringe pattern is as shown in figure.

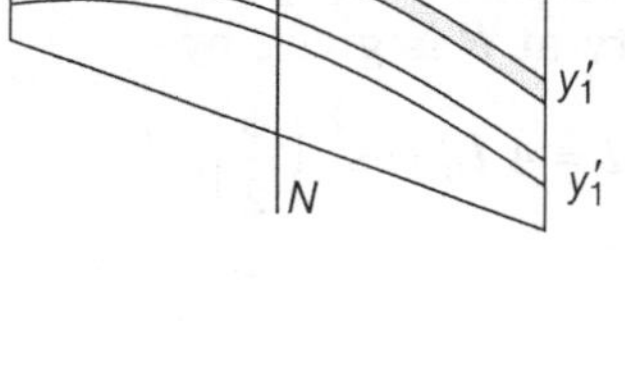

MN is the centre line of screen

on y_0 fringe, $\Delta x = S_1 P - S_2 P = 0$

on y_1 fringe, $\Delta x = S_2 P - S_1 P = \lambda$

on y_1' fringe, $\Delta x = S_1 P - S_2 P = \lambda$

and so on.

Diffraction of Light

- Bending of light at an obstacle or an aperture with size equivalent to the range of wavelength of light is called diffraction of light.
- The diffraction phenomenon is divided into two types:
  Fresnel diffraction and Fraunhofer diffraction.
- In the first type, either source or screen or both are at finite distance from the diffracting device (obstacle or aperture).
- In the second type, both source and screen are effectively at infinite distance from the diffracting device. Fraunhofer diffraction is a particular limiting case of Fresnel diffraction.

Note *Diffraction is a wave property which can also be observed in sound. Since wavelength of sound is large, so diffraction of sound can also be observed when size of aperture is large.*

Single Slit Diffraction

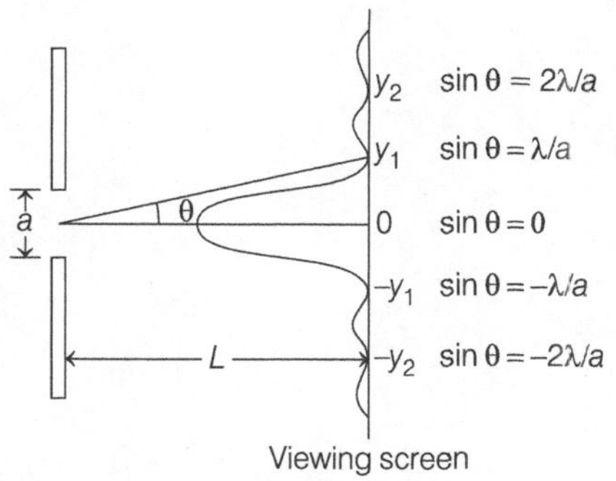

The minimum intensity occurs, where

$$\sin\theta = m\frac{\lambda}{a}, \text{where } m = \pm 1, \pm 2$$

Difference between Interference and Diffraction

Both interference and diffraction are the results of superposition of waves, so they are often present simultaneously as in Young's double slit experiment.

However, interference is the result of superposition of waves from two different wavefronts while diffraction results due to superposition of wavelets from different points of the same wavefronts.

Polarisation

- **Malus' law,**

$$I_2 = I_1 \cos^2 \theta$$

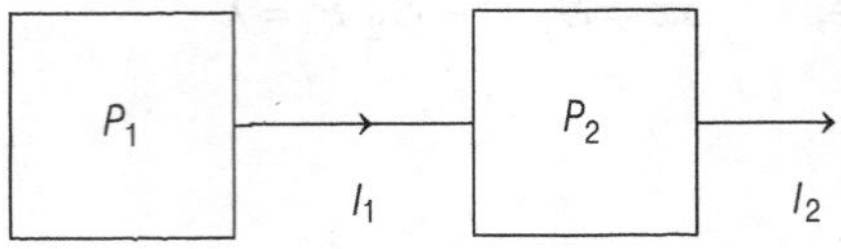

where, θ is the angle between pass axes of two polaroids P_1 and P_2 as shown in above figure.

- **Brewster's law,**

$$\mu = \tan i_p$$

where, i_p is called Brewster's angle.

- At Brewster's angle of incidence, reflected ray and refracted ray are mutually perpendicular.

CHAPTER 17

Electrostatics

Electrostatic Force

- $q = \pm ne, e = 1.6 \times 10^{-19}$C
- Unlike charges attract each other, however like charges repel each other.
- $F = \frac{1}{4\pi\varepsilon_0} \cdot \frac{q_1 q_2}{r^2}$ = magnitude of electrostatic force between two charges.
- In vector form, electrostatic force on charge q_1 due to charge q_2 is given by

$$\mathbf{F}_1 = \frac{1}{4\pi\,\varepsilon_0} \cdot \frac{q_1 q_2}{r^3} (\mathbf{r}_1 - \mathbf{r}_2)$$

In this formula, substitute q_1 and q_2 with sign. Further, r is the distance between two charges and $\mathbf{r}_1$, $\mathbf{r}_2$ are their position vectors.

- In a dielectric (insulating) medium, force decreases K times, where K is the dielectric constant of that medium.

Hence, $$F = \frac{F_0}{K} = \frac{1}{4\pi\,\varepsilon_0 K} \cdot \frac{q_1 q_2}{r^2}$$

Here, F_0 = force between the charges in vacuum.

Note *In some problems of electrostatics, Lami's theorem is very useful. According to this theorem, "if three concurrent forces* $\mathbf{F}_1$, $\mathbf{F}_2$ *and* $\mathbf{F}_3$ *as shown in figure are in equilibrium or if* $\mathbf{F}_1 + \mathbf{F}_2 + \mathbf{F}_3 = 0$, *then*

$$\frac{F_1}{\sin\alpha} = \frac{F_2}{\sin\beta} = \frac{F_3}{\sin\gamma}$$

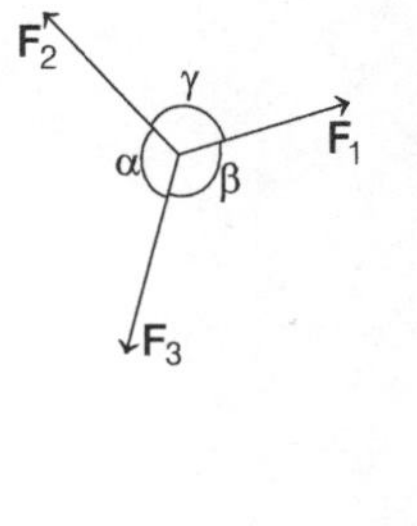

Electric Field and Potential

- Field strength due to a point charge,

$$E = \frac{1}{4\pi\varepsilon_0} \cdot \frac{q}{r^2}$$

- Due to positive charge, field strength is away from the charge and due to negative charge, it is towards the charge.

- In vector form, field strength at point P is given by

$$\mathbf{E}_P = \frac{1}{4\pi\varepsilon_0} \cdot \frac{q}{r^3} (\mathbf{r}_p - \mathbf{r}_q)$$

In this formula substitute q with sign.

Here, $\mathbf{r}_q$ is the position vector of point where q is kept.

- $\mathbf{F} = q\mathbf{E}$. On positive charge, electrostatic force is qE in the direction of electric field. Force on negative charge is also qE in the opposite direction of electric field.
- Electric potential (V) at some point is defined as the negative of work done by electrostatic forces in bringing unit positive test charge from infinity to that point.

Due to a point charge, $V = \frac{1}{4\pi\,\varepsilon_0}\frac{q}{r}$, substitute q with sign.

- **Principle of superposition** At some point net field strength due to more than one point charges is the vector sum due to individual point charges and net potential is the scalar sum.
- **List of formulae for field strength E and potential V**

Charge Distribution	E		V	
	Formula	Graph	Formula	Graph
Uniformly charged spherical shell	$E_i = 0$ $E_s = K\cdot\frac{q}{R^2} = \frac{\sigma}{\varepsilon_0}$ $E_o = \frac{Kq}{r^2}$	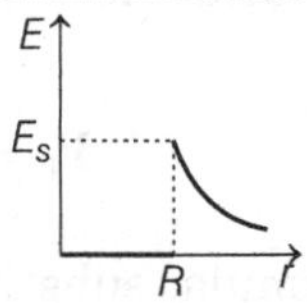	$V_i = V_s = \frac{Kq}{R}$ $= \frac{\sigma R}{\varepsilon_0}$ $V_o = \frac{Kq}{r}$	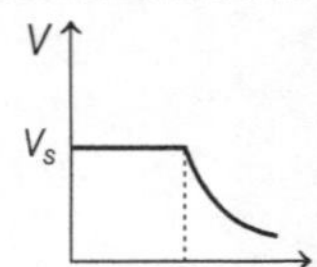
Solid sphere of charge	$E_i = \frac{Kqr}{R^3}$ $E_s = \frac{Kq}{R^2}$ $E_o = \frac{Kq}{r^2}$	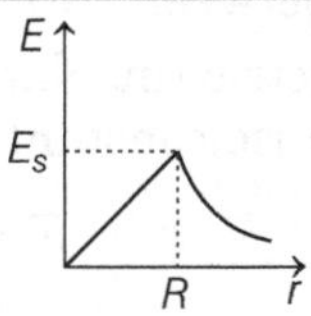	$V_i = \frac{Kq}{R^3}$ $(1.5R^2 - 0.5r^2)$ $V_s = \frac{Kq}{R}, V_o = \frac{Kq}{r}$	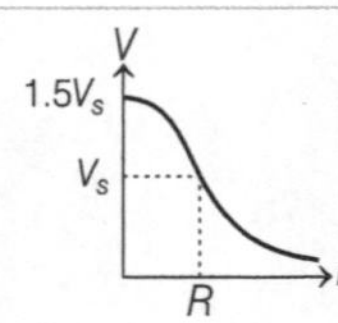
On the axis of a uniformly charged ring	$E = \frac{Kqx}{(R^2 + x^2)^{3/2}}$ At centre $x = 0$ $\therefore E = 0$	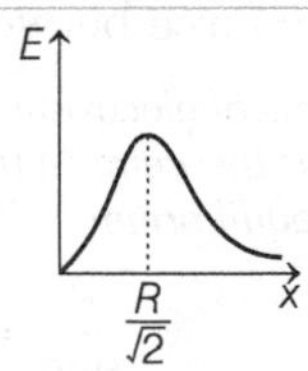	$V = \frac{Kq}{\sqrt{R^2 + x^2}}$ At centre $x = 0$ $\therefore V = \frac{Kq}{R}$	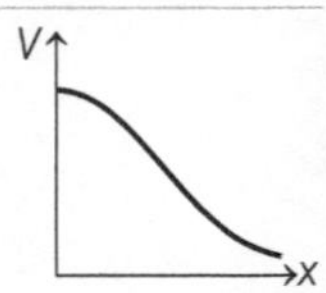
Infinitely long line charge	$E = \frac{\lambda}{2\pi\,\varepsilon_0 r}$	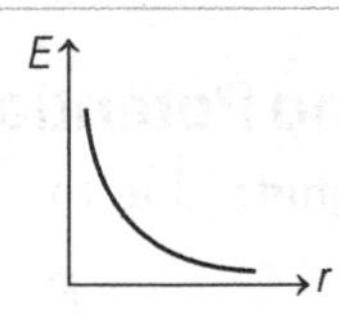	$P.D = \frac{\lambda}{2\pi\,\varepsilon_0}\ln\left(\frac{r_2}{r_1}\right)$	Not required
Infinitely long thin sheet of charge	$E = \frac{\sigma}{2\varepsilon_0}$	Electric field is uniform	$P.D = Ed$	Not required

- Potential difference between two charged spherical shells is given by

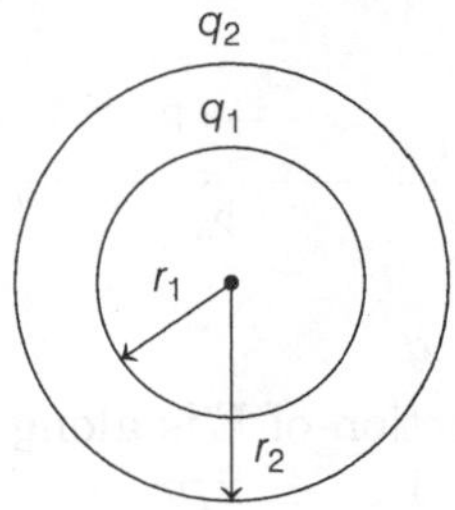

$$V_1 - V_2 = \frac{q_1}{4\pi\varepsilon_0}\left(\frac{1}{r_1} - \frac{1}{r_2}\right)$$

- Potential difference depends only upon the charge on inner shell q_1. When the two charged spherical shells are connected by a conducting wire, whole charge q_1 transfers to outer shell and potential difference between two shells becomes zero.

Relation between *E* and *V*

- $\mathbf{E} = -\left[\frac{\partial V}{\partial x}\hat{\mathbf{i}} + \frac{\partial V}{\partial y}\hat{\mathbf{j}} + \frac{\partial V}{\partial z}\hat{\mathbf{k}}\right]$
- $E = -\frac{dV}{dr}$ or $-\frac{dV}{dx} = -$ slope of V-r or V-x
- $V_b - V_a = -\int_a^b \mathbf{E}\cdot d\mathbf{r}$ as $dV = -\mathbf{E}\cdot d\mathbf{r}$

  Here, $d\mathbf{r} = dx\hat{\mathbf{i}} + dy\hat{\mathbf{j}} + dz\hat{\mathbf{k}}$

  In case of one variable we can write, $dV = -Edr$ or $-Edx$
- In a uniform electric field, potential difference, $V = Ed$

  To check whether which point is at higher potential and which point is at lower potential, we will use the fact that electric lines always flow from higher potential to lower potential.

Electrostatic Potential Energy (*U*)

- Electrostatic potential energy is the negative of work done by electrostatic forces in making a system of charges from infinity to the present position.
- Electrostatic potential energy of two point charges, $U = \frac{1}{4\pi\varepsilon_0}\cdot\frac{q_1 q_2}{r}$

  In this formula, substitute q_1 and q_2 with sign.
- For a system of charges of more than two point charges, pairs are formed.

  For example, potential energy of four charges will be

$$U = \frac{1}{4\pi\varepsilon_0}\left[\frac{q_4 q_3}{r_{43}} + \frac{q_4 q_2}{r_{42}} + \frac{q_4 q_1}{r_{41}} + \frac{q_3 q_2}{r_{32}} + \frac{q_3 q_1}{r_{31}} + \frac{q_2 q_1}{r_{21}}\right]$$

  If there are n point charges, then total number of pairs are $\frac{n(n-1)}{2}$.

- $U = qV$

Electric dipole

- $p = q\,(2a)$

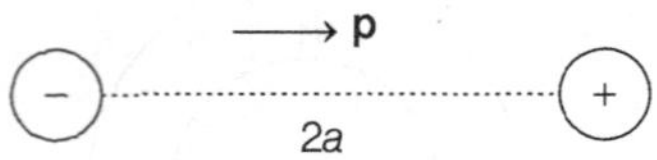

- Direction of **p** is from $-q$ to $+q$.
- On the axis of the dipole, direction of **E** is along **p** and given by

$$E = \frac{1}{4\pi\varepsilon_0} \cdot \frac{2pr}{(r^2 - a^2)^2}$$

$$\approx \frac{1}{4\pi\varepsilon_0} \cdot \frac{2p}{r^3} \qquad \text{(for } r >> a\text{)} \qquad \left(\text{or } E \propto \frac{1}{r^3}\right)$$

- On the axis of dipole, potential is given by

$$V = \frac{1}{4\pi\varepsilon_0} \frac{p}{r^2 - a^2}$$

$$\approx \frac{1}{4\pi\,\varepsilon_0} \frac{p}{r^2} \text{ for } r >> a$$

- On perpendicular bisector of the dipole, **E** is antiparallel to **p** and is given by

$$E = \frac{1}{4\pi\varepsilon_0} \frac{p}{(r^2 + a^2)^{3/2}} \approx \frac{1}{4\pi\varepsilon_0} \frac{p}{r^3} \qquad \text{(for } r >> a\text{)}$$

Electric dipole when placed in uniform electric field

- $\mathbf{F} = 0$
- $\tau = \mathbf{p} \times \mathbf{E}$ or $\tau = pE \sin\theta$
- $U = -\mathbf{p} \cdot \mathbf{E} = -\,pE\cos\theta$
- $W_{\theta_1 \to \theta_2} = U_{\theta_2} - U_{\theta_1} = pE\,(\cos\theta_1 - \cos\theta_2)$
- $\theta = 0°$ is stable equilibrium position.
  In this position, $F = 0$, $\tau = 0$, $U = -\,pE$ = minimum.
- $\theta = 180°$ is unstable equilibrium position.
  In this position, $F = 0$, $\tau = 0$, $U = +\,pE$ = maximum.
  In the above formulae θ is the angle between **p** and **E**.

Electric Flux and Gauss's law

- Electric flux from a surface, $\phi = \int \mathbf{E} \cdot d\mathbf{S}$
- Gauss's law, $\oint \mathbf{E} \cdot d\mathbf{S} = \frac{q_{in}}{\varepsilon_0}$ or total electric flux from a closed surface $= \frac{q_{in}}{\varepsilon_0}$

Properties of a conductor

- Excess charge on a conductor resides on its outer surface.
- Under electrostatic conditions,

$$E_i = 0 \quad \text{and} \quad E_s = \frac{\sigma}{\varepsilon_0}$$

- On the surface of charged conductor, electric lines are perpendicular to the surface.
- The potential of a charged conductor throughout its volume is same.

Electric field lines

- The tangent to a line at any point gives the direction of **E** at that point. This is also the direction of electrostatic force on positive charge at that point.
- Electric field lines always begin from a positive charge and end on a negative charge and do not start or stop in mid-space.
- Two lines can never intersect. If it happens, then two tangents can be drawn at their point of intersection, i.e. intensity at that point will have two directions which is absurd.
- In a uniform field, the field lines are straight parallel and uniformly spaced.

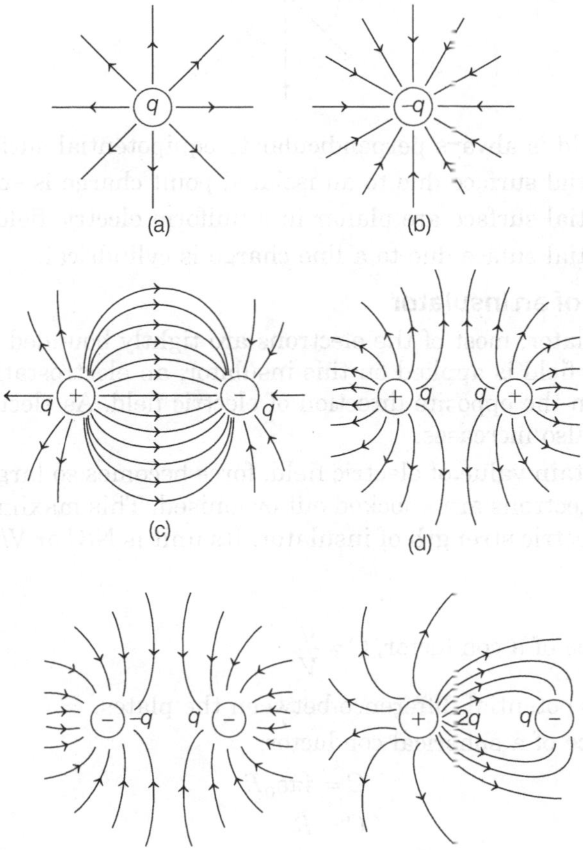

- The electric field lines can never form closed loops as a field line can never start and end on the same charge.

- Electric field lines always flow from higher potential to lower potential.
- In a region where, there is no electric field, lines are absent. This is why inside a conductor (where, electric field is zero), there cannot be any electric field line.
- Electric lines of force ends or starts normally from the surface of a conductor.

Equipotential surface

- Equipotential surface is an imaginary surface joining the points of same potential in an electric field. So, we can say that the potential difference between any two points on an equipotential surface is zero.
- The electric lines of force at each point of an equipotential surface are normal to the surface.

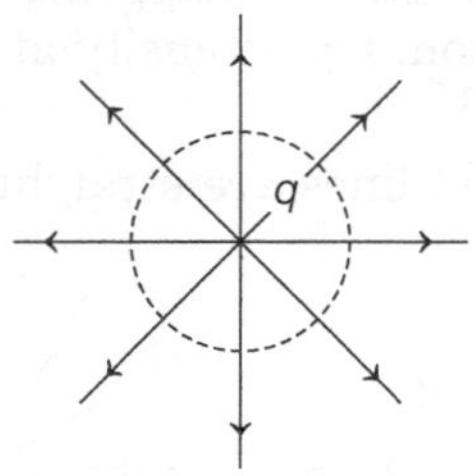

- Electric field is always perpendicular to equipotential surface.
- Equipotential surface due to an isolated point charge is spherical.
- Equipotential surface are planar in a uniform electric field.
- Equipotential suface due to a line charge is cylindrical.

Dielectric strength of an insulator

- In an insulator, most of the electrons are tightly bounded with the nucleus. If an electric field is applied on this insulator, an electrostatic force acts on these electrons in the opposite direction of electric field. As electric field increases, this force also increases.

  After a certain value of electric field, force becomes so large that these bounded electrons are knocked out or ionised. This maximum electric field is called dielectric strength of insulator. Its unit is N/C or V/m.

Capacitors

- Capacitance of a conductor, $C = \frac{q}{V}$

  where, V = potential difference between the plates.
- Capacitance of a spherical conductor,

$$C = 4\pi\varepsilon_0 R$$

or

$$C \propto R$$

or

$$\frac{C_1}{C_2} = \frac{R_1}{R_2}$$

- Capacitance of parallel plate capacitor

$$C = \frac{\varepsilon_0 A}{d} \quad \text{(vacuum between two plates)}$$

- When a dielectric slab is completely filled between the plates,

$$C = \frac{K\varepsilon_0 A}{d}$$

- When a dielectric slab is partially filled between the plates,

$$C = \frac{\varepsilon_0 A}{d - t + \frac{t}{K}}$$

- Potential energy stored in the electric field of a charged conductor or capacitor,

$$U = \frac{1}{2}CV^2 = \frac{1}{2}\frac{q^2}{C} = \frac{1}{2}qV$$

- Energy density, $u = \frac{1}{2}\varepsilon_0 E^2$
- **Capacitors in series**
  (i) Charge on capacitors is same.
  (ii) $\frac{1}{C} = \frac{1}{C_1} + \frac{1}{C_2} + \dots + \frac{1}{C_n}$
  (iii) Potential difference distributes in inverse ratio of capacity.
- **Capacitors in parallel**
  (i) Potential difference is same.
  (ii) $C = C_1 + C_2 + \dots + C_n$
  (iii) Charge distributes in direct ratio of capacity.
- Capacitance of spherical capacitor,

$$C = \frac{4\pi\varepsilon_0}{\left(\frac{1}{a} - \frac{1}{b}\right)} = 4\pi\varepsilon_0\left(\frac{ab}{b-a}\right)$$

- **Redistribution of charge** When two isolated charged conductors are connected to each other, then total charge is redistributed in the ratio of their capacitances.
  (i) Common potential, $V = \frac{q_1 + q_2}{C_1 + C_2} = \frac{C_1V_1 + C_2V_2}{C_1 + C_2}$
  (ii) Energy loss $= \frac{1}{2}\frac{C_1C_2(V_1 - V_2)^2}{(C_1 + C_2)}$

CHAPTER 18

Current Electricity

Electric Current and Resistance

- Current in a simple circuit $i = \dfrac{\text{net emf}}{\text{net resistance}}$.

  A simple circuit means a single wire problem.
- Current density $J = \dfrac{i}{A}$
- $i = neAv_d$, where v_d = drift velocity.
- Flow of charge

$$\Delta q = i\Delta t \quad \text{(If current is constant)}$$
$$= \int i\,dt \quad \text{(If current is a function of time)}$$
$$= \text{area under } i\text{ - }t \text{ graph}$$

- $R = \rho\dfrac{l}{A}$, where ρ = specific resistance $= \dfrac{1}{\sigma}$
- $R_t = R_0\ [1 + \alpha\,(t - t_0)]$
- **Series combination of resistors,**

$$R = R_1 + R_2$$

In series, potential difference distributes in the direct ratio of resistance, or

$$V \propto R \Rightarrow \frac{V_1}{V_2} = \frac{R_1}{R_2}$$

- **Parallel combination of resistors,**

$$\frac{1}{R} = \frac{1}{R_1} + \frac{1}{R_2} \quad \text{or} \quad R = \frac{R_1R_2}{R_1 + R_2}$$

In parallel, current distributes in the inverse ratio of resistance

or $$i \propto \frac{1}{R} \Rightarrow \frac{i_1}{i_2} = \frac{R_2}{R_1} \quad \text{or} \quad i_1 : i_2 : i_3 = \frac{1}{R_1} : \frac{1}{R_2} : \frac{1}{R_3}$$

In parallel grouping of resistors, net resistance is less than the smallest resistance.

Thermal Motion and Drift Motion

- Free electrons inside a solid conductor can have two motions:
  (i) Random or thermal motion (speed of the order of 10^5 m/s)
  (ii) Drift motion (speed of the order of 10^{-4} m/s)
- Net current due to random (or thermal motion) is zero from any section, whereas net current due to drift motion is non-zero.
- In the absence of any electric field (or a potential difference across the conductor) free electrons have only random motion. Hence, net current from any section is zero.
- In the presence of an electric field (or a potential difference across the conductor) free electrons have both motions (random and drift). Therefore, current is non-zero due to drift motion.

Relaxation Time

- Relaxation time τ is the average time between two successive collisions. Its value is of the order of 10^{-14} s.

$$v_d = \frac{eE\tau}{m}$$

- If a current i is flowing through a wire of non-uniform cross-section, then current will remain constant at all cross-sections.

But drift velocity and current density are inversely proportional to the area of cross-section. This is because

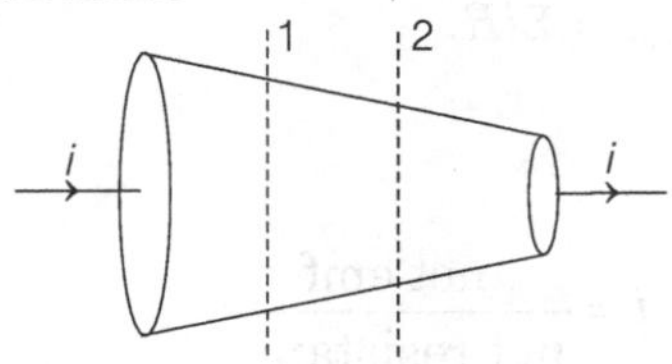

$$i = neAv_d \quad \text{or} \quad v_d = \frac{i}{neA} \quad \text{or} \quad v_d \propto \frac{1}{A}$$

Further, $$j = \frac{i}{A} \quad \text{or} \quad j \propto \frac{1}{A}$$

So, in the figure $i_1 = i_2 = i$, but $(v_d)_2 > (v_d)_1$ and $j_2 > j_1$ because $A_2 < A_1$

Ohm's Law

- According to Ohm's law, there are some of the materials (like metals or conductors) or some circuits for which, current passing through them is proportional to the potential difference applied across them.

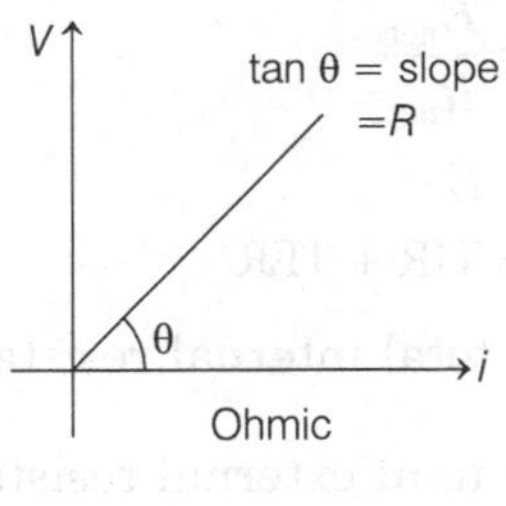

or $i \propto V$

or $V \propto i$

$\Rightarrow$ $V = iR$

or $\frac{V}{i} = R = \text{constant}$

or V - i graph for such materials and circuits is a straight line passing through origin. Slope of this graph is called its resistance.

- The materials or circuits which follow this law are called ohmic materials.
- The materials or circuits which do not follow this law are called non-ohmic materials. V - i graph for non-ohmic circuits is not a straight line passing through origin.

Kirchhoff's Law

- Current in a complex circuit (more than one wire problems) can be obtained with the help of Kirchhoff's two laws.
- **Kirchhoff's first law** is applied at a junction. It is law of conservation of charge. It is given by

$$\Sigma i = 0$$

or Net incoming current = net outgoing current.

- **Kirchhoff's second law** is applied in a closed loop. It is law of conservation of energy. It is given by

$$\Sigma E_{\text{net}} = \Sigma iR$$

Grouping of Batteries

- **Series grouping**

$$i = \frac{\text{net emf}}{\text{net resistance}}$$

If there are n identical batteries of emf E and internal resistance r, then current through external resistance R (if all batteries are additive in nature) is given by

$$i = \frac{nE}{nr + R}$$

If polarity of m batteries is reversed, then

$$i = \frac{(n - 2m)E}{nr + R}$$

- **Parallel grouping with identical batteries**

$$i = \frac{E_{\text{net}}}{R_{\text{net}}}$$

Here, $E_{\text{net}} = E$

and $R_{\text{net}} = \text{TIR} + \text{TER}$

where, TIR = total internal resistance $= \frac{r}{n}$

TER = total external resistance $= R$

- **Parallel grouping with unidentical batteries**

$$i = \frac{E_{net}}{R_{net}}$$

Here, $$E_{net} = \frac{\Sigma (E/r)}{\Sigma (1/r)} \quad \text{...(i)}$$

If any of the battery is oppositely connected, then place negative sign in numerator of Eq. (i), but no change in denominator.

- **Mixed grouping**

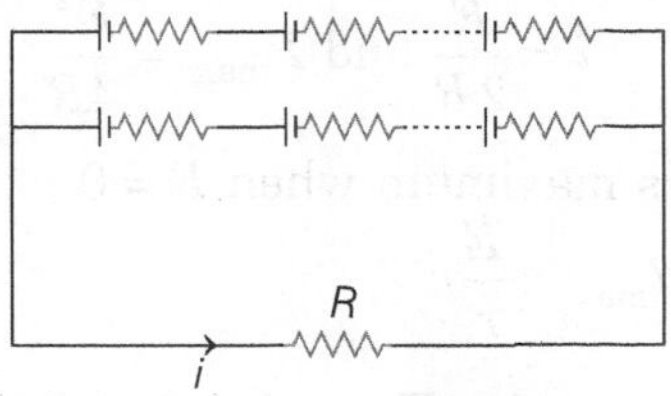

If there are n rows of identical batteries (E, r) with m cells in each row, then

$$i = \frac{E_{net}}{R_{net}}$$

Here, $$E_{net} = mE$$

$$R_{net} = \text{TIR} + \text{TER}$$

$$\text{TIR} = \text{total internal resistance} = \frac{mr}{n}$$

$$\text{TER} = \text{total external resistance} = R$$

In this case, current through the external resistance is maximum when

$$\text{TIR} = \text{TER}$$

or $$R = \frac{mr}{n}$$

Power and Heat Generated Across a Resistance

- $$P = \frac{V^2}{R} = i^2R = Vi$$
- $$H = \frac{V^2}{R}t = i^2Rt = Vit$$

Here, V is potential difference across R and i is the current through that resistance R only.

- In the circuit shown in figure, if r is variable, then current through R or power across R is maximum when

$$r = 0$$

Hence, $$i_{max} = \frac{E}{R} \quad \text{and} \quad P_{max} = \frac{E^2}{R}$$

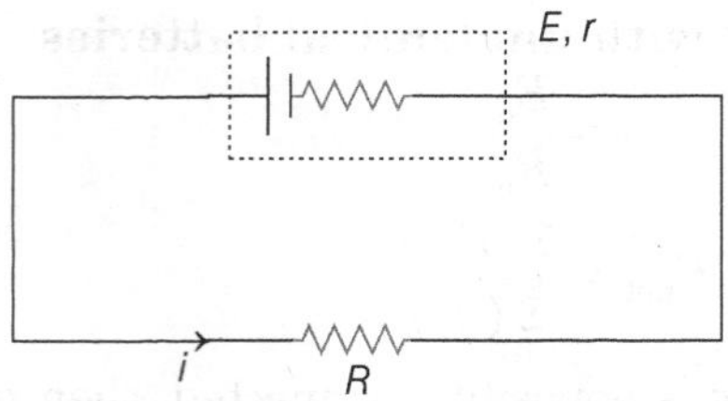

If R is variable, power is maximum, when $R = r$ or total external resistance = total internal resistance.

In this case, $i = \dfrac{E}{2R}$ and $P_{max} = \dfrac{E^2}{4R}$

In this case, current is maximum when $R = 0$.

$$i_{max} = \frac{E}{r}$$

Potential Difference Across the Terminals of a Battery

- $V = E$, if $i = 0$
- $V = 0$, if battery is short circuited.
- $V = E - ir$, if battery is supplying the energy or current is flowing in the normal direction as shown below.

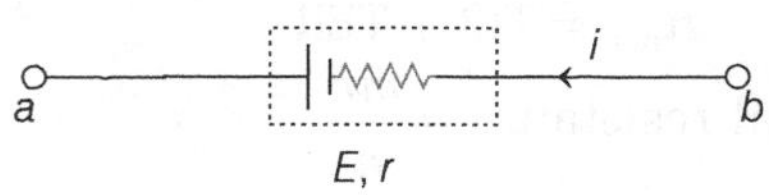

$$V = V_a - V_b = V_{ab} = E - ir$$

- $V = E + ir$, if battery is being charged or current is flowing in opposite direction as shown below.

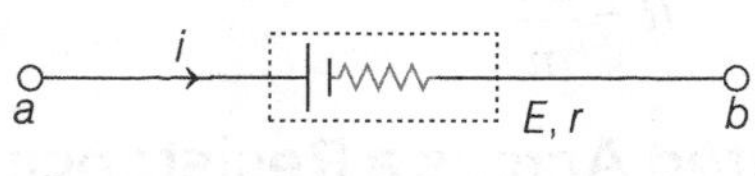

$$V = V_a - V_b = V_{ab} = E + ir$$

Conversion of Galvanometer into Ammeter

A galvanometer is converted into an ammeter by connecting a low resistance (called shunt) in parallel with galvanometer. This assembly (called ammeter) is connected in series in the wire in which current is to be measured.

Resistance of an ideal ammeter should be zero.

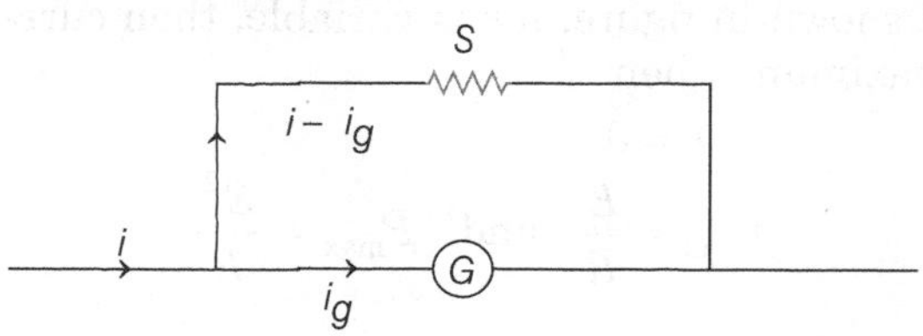

$$\frac{S}{G} = \frac{i_g}{i - i_g}$$

$$\therefore \quad S = \text{shunt} = \left(\frac{i_g}{i - i_g}\right) G$$

Conversion of Galvanometer into Voltmeter

A galvanometer is converted into a voltmeter by connecting a high resistance in series with galvanometer. The whole assembly called voltmeter is connected in parallel between two points across which potential difference is to be found. Resistance of an ideal voltmeter should be infinite.

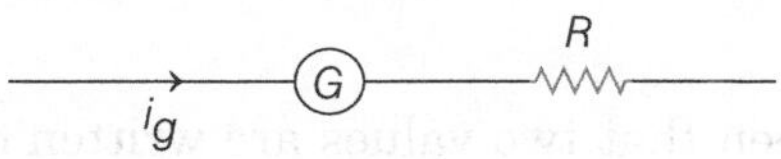

$$V = i_g\,(G + R)$$

$$\therefore \quad R = \text{high resistance required in series} = \frac{V}{i_g} - G$$

Principle of Potentiometer

$$V_{ac} = V_{de}$$

or $$i_1 \lambda\, l = E_2 - i_2 r \qquad \ldots\text{(i)}$$

Here, λ is resistance per unit length of potentiometer wire ab.

If $i_2 = 0$, then $i_1 \lambda\, l = E_2$

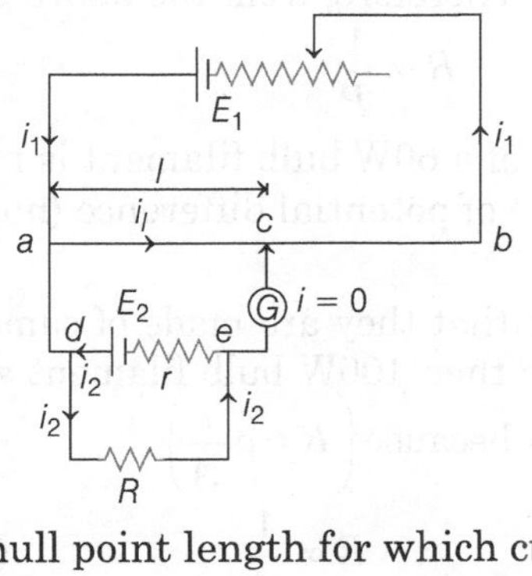

In Eq. (i), l is called null point length for which current through galvanometer is zero.

$$i_1 = \frac{E_1}{R_{ab} + R_h}$$

Here, R_{ab} = resistance of potentiometer wire ab,

R_h = resistance of rheostat.

and $$i_2 = \frac{E_2}{R + r}$$

- Internal resistance of a battery in the experiment of potentiometer

$$r = R\left(\frac{E}{V} - 1\right) = R\left(\frac{l_1}{l_2} - 1\right)$$

Here, l_1 = null point length when $i_2 = 0$ and
l_2 = null point length when $i_2 \neq 0$.

- Comparison of EMF's of two batteries in the experiment of potentiometer,

$$\frac{E_1}{E_2} = \frac{l_1}{l_2}$$

Here, l_1 and l_2 are null point lengths for $i_2 = 0$, corresponding to two unknown batteries.

Principle of a Bulb

- You might have seen that two values are written over a bulb, power and potential difference. From the written values, we can find resistance of the filament of bulb by the equation,

$$P = \frac{V^2}{R}$$

or

$$R = \frac{V^2}{P}$$

For example, on a 60 W bulb you will see two values written over it, 60 W and 220 V. It implies that, if a potential difference of 220 V is applied on this bulb, then it will consume 60 J of electrical power in 1 s or we can say, it will consume 60W of electrical power.

- In two different bulbs normally written value of V is same but written value of P will be different. Therefore, from the above equation, we can see that

$$R \propto \frac{1}{P} \qquad \text{(As } V \text{ is same)}$$

Therefore, resistance of a 60W bulb filament is more than the 100W bulb filament if same value of potential difference (normally 220V) is written over them.

- Further if we assume that they are made of same material and their filament lengths are also same then 100W bulb filament will be thicker than the 60W bulb filament. This is because $\left(R = \rho\frac{l}{A}\right)$

$$R \propto \frac{1}{A} \qquad \text{(As } \rho \text{ and } l \text{ are assumed same)}$$

$$R_{100\text{W}} < R_{60\text{W}}$$

$$\therefore \quad A_{100\text{W}} > A_{60\text{W}}$$

- Now, actual potential difference across the bulb may be greater than or less than the written value of potential difference. Accordingly actual power consumed by the bulb may be greater than or less than the written value of power.

For example, a (60W, 220V) bulb will consume more than 60W power, if actual applied potential difference is more than 220V and it will consume less than 60W power, if actual applied potential difference is less than 220V.

- Some more points are as under :

  (i) Resistance of bulb, $R = \dfrac{V^2}{P}$ or $R \propto \dfrac{1}{P}$

  (ii) In parallel, $P = P_1 + P_2$

  (iii) In series, $\dfrac{1}{P} = \dfrac{1}{P_1} + \dfrac{1}{P_2}$ or $P = \dfrac{P_1 P_2}{P_1 + P_2}$

- In $R = \dfrac{V^2}{P}$, V and P are rated (written) values for that bulb. In parallel, a bulb having more rated power glows more brightly. In series, a bulb having less rated power glows more brightly.
- In series, since i is same, hence

$$P \propto R \quad \text{or} \quad \frac{P_1}{P_2} = \frac{R_1}{R_2}$$

In parallel, since V is same, hence,

$$P \propto \frac{1}{R}$$

or

$$\frac{P_1}{P_2} = \frac{R_2}{R_1}$$

- Power supplied/consumed by a battery $= \pm Ei$
- EMF of a battery, $E = \dfrac{W}{q}$

Short Circuiting

Two points in an electric circuit directly connected by a conducting wire are called short-circuited. Under such condition, both points are at same potential.

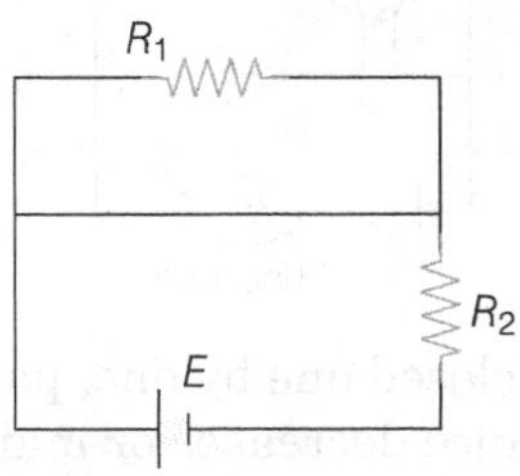

For example, resistance R_1 in the adjoining circuit is short circuited, i.e. potential difference across it is zero. Hence, no current will flow through R_1 and the current through R_2 is therefore, E/R_2.

Earthing

If some point of a circuit is earthed, then its potential is taken to be zero. For example, in the below figure

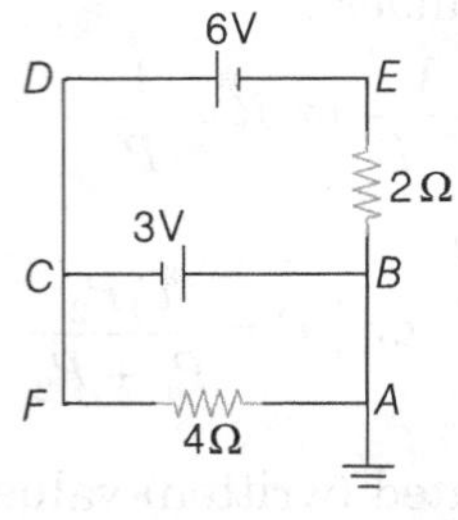

$$V_A = V_B = 0$$
$$V_F = V_C = V_D = -3\text{ V}$$
$$V_E = -9\text{ V}$$

$\therefore$ $$V_B - V_E = 9\text{ V}$$

or current through 2 Ω resistance is

$$\frac{V_B - V_E}{2} \text{ or } \frac{9}{2}\text{ A} \qquad \text{(From } B \text{ to } E\text{)}$$

Similarly, $$V_A - V_F = 3\text{ V}$$

and the current through 4 Ω resistance is $\frac{V_A - V_F}{4}$ or $\frac{3}{4}$ A (From A to F)

Fuse Wire

Different electrical devices (like bulbs, fans, coolers, etc.) in our houses are connected in parallel with main line. A fuse wire is connected with main current i.

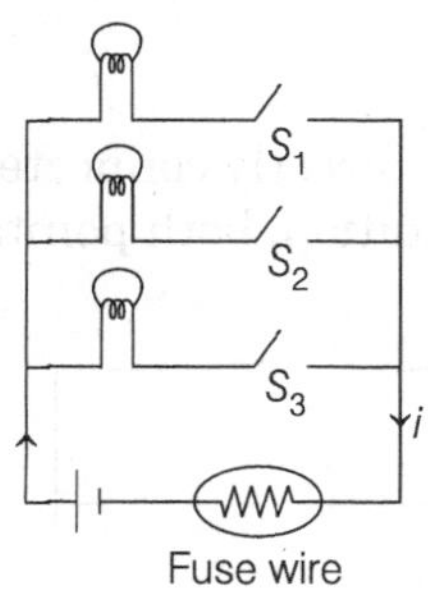

When different switches are closed one by one, parallel resistors increase in circuit. Therefore, net resistance decreases or main current i increases with increase in the value of i, heat developed across the fuse wire ($H = i^2Rt$) also increases.

After a certain maximum value, fuse wire melts and main circuit is disconnected.

Fuse wire is generally prepared from tin-lead alloy. It should have high resistivity and moderate melting point (not too high, not too low).

Kilowatt Hour

The commercial unit of electric energy is called kilowatt-hour (kWh) or Board of Trade Unit (BOT) or simply unit.

If an electrical appliance of 1 kilowatt (= 1000W) is run for 1 hour, then it will consume 1 kilowatt-hour electric energy.

Thus, $1\text{ kWh} = (1\text{ kilowatt}) \times (1\text{ hour})$

$$= (1000\text{ W})(1\text{ h})$$
$$= (1000\text{ J/s})(60 \times 60\text{ s})$$
$$= 3.6 \times 10^6\text{ J}$$

or $$1\text{ kWh} = 3.6 \times 10^6\text{ J}$$

Wheatstone Bridge

If $\frac{P}{Q} = \frac{R}{S}$, then bridge is said to be balanced. In balanced condition,

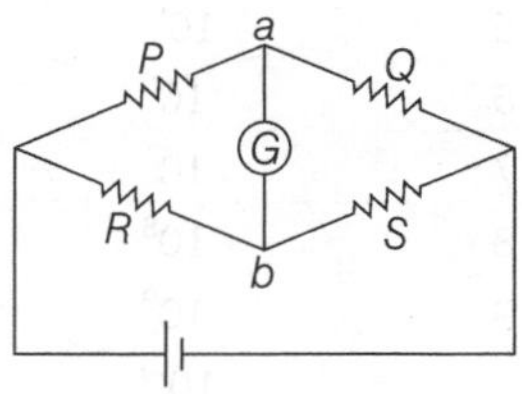

$$V_a = V_b$$

or $$V_{ab} = 0$$

or $$i_G = 0$$

Meter Bridge

This is based on balanced Wheatstone bridge. This is usually to find some unknown resistance say X.

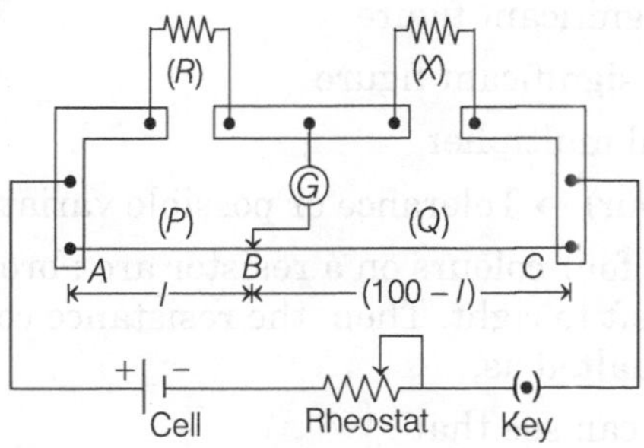

When $I_G = 0$, bridge is balanced and

$$\frac{P}{Q} = \frac{R}{X}$$

$$\therefore \quad \frac{l}{100 - l} = \frac{R}{X}$$

or $$X = \left(\frac{100 - l}{l}\right) R$$

Most accurate results are obtained when null point (when $i_G = 0$) is obtained near the centre of wire AC or the ratio $\frac{P}{Q}$ or $\frac{R}{X}$ is kept 1 : 1.

Colour Codes for Resistors

Colour	Number	Multiplier	Tolerance (%)
Black	0	1	
Brown	1	10^1	
Red	2	10^2	
Orange	3	10^3	
Yellow	4	10^4	
Green	5	10^5	
Blue	6	10^6	
Violet	7	10^7	
Gray	8	10^8	
White	9	10^9	
Gold		10^{-1}	5
Silver		10^{-2}	10
No colour			20

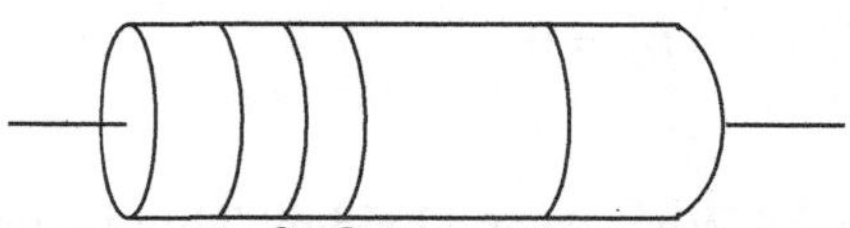

Colour 1 → First significant figure
Colour 2 → Second significant figure
Colour 3 → Decimal multiplier
Colour 4 (or no colour) → Tolerance or possible variation in percentage.

For example The four colours on a resistor are : brown, yellow, green and gold as read from left to right. Then, the resistance corresponding to these colours can be calcualted as,

From the table, we can see that

Brown colour → 1
Yellow colour → 4
Green colour → 10^5

and Gold colour → 5 %

$\therefore$ $R = (14 \times 10^5 \pm 5\%)\,\Omega$

CHAPTER 19

Magnetic Effects of Current & Magnetism

Force on a Charged Particle in a Uniform Magnetic Field

- **Magnetic force on a moving charge**

$$\mathbf{F}_m = q\,(\mathbf{v} \times \mathbf{B})$$

or
$$F_m = Bqv\sin\theta$$

- Path of charged particle in a uniform magnetic field
  (i) If $\theta = 0°$ or $180°$, path is straight line.
  (ii) If $\theta = 90°$, path is circle.
  (iii) For any other angle path is helix.

Note *In the above formulae, θ is the angle between* **v** *and* **B**.

- **List of formulae in circular path**
  (i) $r = \dfrac{mv}{Bq} = \dfrac{p}{Bq} = \dfrac{\sqrt{2Km}}{Bq} = \dfrac{\sqrt{2qVm}}{Bq}$
  (ii) $T = \dfrac{2\pi m}{Bq}$; $\omega = \dfrac{Bq}{m}$; $f = \dfrac{Bq}{2\pi m}$

- **List of formulae in helical path**
  (i) $r = \dfrac{mv\sin\theta}{Bq}$
  (ii) $T = \dfrac{2\pi m}{Bq}$; $f = \dfrac{Bq}{2\pi m}$
  (iii) Pitch of helical path, $p = (v\cos\theta)\,T = \dfrac{2\pi mv\cos\theta}{Bq}$

- Path of a charged particle in uniform electric and magnetic field will remain unchanged, if

$$\mathbf{F}_{\text{net}} = 0$$

or $$\mathbf{F}_e + \mathbf{F}_m = 0$$
or $$q\mathbf{E} + q\,(\mathbf{v} \times \mathbf{B}) = 0$$
or $$\mathbf{E} = -(\mathbf{v} \times \mathbf{B}) = (\mathbf{B} \times \mathbf{v})$$

Force on Current Carrying Conductor

- Magnetic force on a straight current carrying conductor
$$\mathbf{F}_m = i\,(\mathbf{l} \times \mathbf{B}) \quad \text{or} \quad F_m = ilB\sin\theta$$
Here, θ is the angle between $\mathbf{l}$ and $\mathbf{B}$ or between direction of current and magnetic field.
- For the magnetic force on an arbitrarily shaped wire segment, let us consider the magnetic force exerted on a small segment of vector length $d\mathbf{l}$.

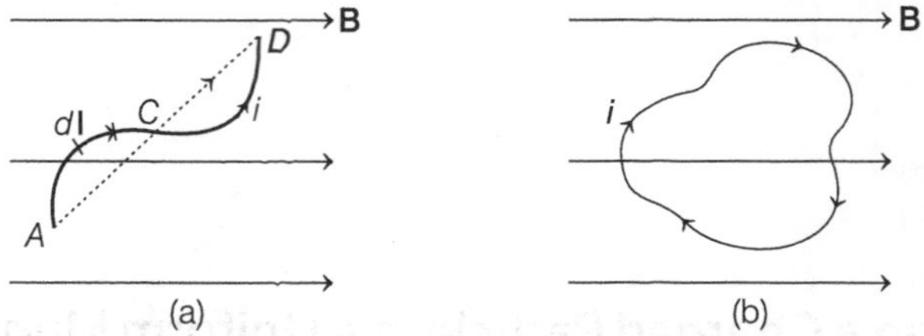

$$\mathbf{F}_m = i\int_A^D (d\mathbf{l} \times \mathbf{B})$$
$$\therefore \quad \mathbf{F}_m = i\left(\int_A^D d\mathbf{l}\right) \times \mathbf{B}$$
But the quantity $\int_A^D d\mathbf{l}$ represents the vector sum of all length elements from A to D. From the polygon law of vector addition, the sum equals the vector $\mathbf{l}$ directed from A to D. Thus,
we can write $\mathbf{F}_{ACD} = \mathbf{F}_{AD} = i\,(\mathbf{AD} \times \mathbf{B})$ in uniform field.
Using the above result ,we can see that net megnetic force in current carrying loop in uniform magnetic field is zero.
- Force per unit length between two parallel current carrying wires,
$$\frac{F}{l} = \frac{\mu_0}{2\pi}\frac{i_1 i_2}{r}$$
There will be attraction between the wires, if currents are in the same direction and repulsion, if currents are in opposite directions.

Magnetic Dipole

- Every current carrying loop is a magnetic dipole.
- Magnetic dipole moment of magnetic dipole is given by
$$M = NiA$$
- Direction of $\mathbf{M}$ is perpendicular to the plane of the loop and given by right hand screw law.
In addition to the method discussed above for finding $\mathbf{M}$, here are two more methods for calculating $\mathbf{M}$.

Method 1 This method is useful for calculating **M** for a rectangular or square loop.

The magnetic moment (**M**) of the rectangular loop shown in figure is

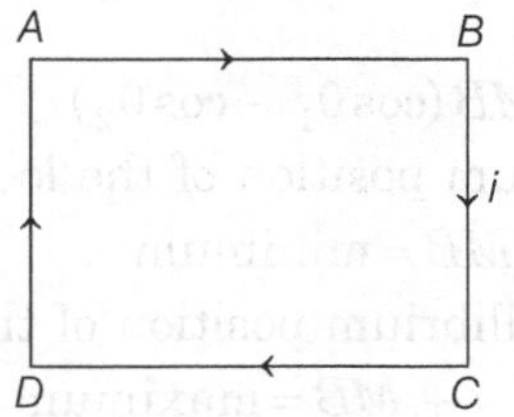

$\mathbf{M} = i\,(\mathbf{AB} \times \mathbf{BC}) = i\,(\mathbf{BC} \times \mathbf{CD}) = i\,(\mathbf{CD} \times \mathbf{DA}) = i\,(\mathbf{DA} \times \mathbf{AB})$

Here, the cross product of any two consecutive sides (taken in order) gives the area as well as the correct direction of **M**.

Method 2 Sometimes, a current carrying loop does not lie in a single plane. By assuming two equal and opposite currents in one branch (which obviously makes no change in the given circuit) two (or more) closed loops are completed in different planes.

Now, the net magnetic moment of the given loop is the vector sum of individual loops.

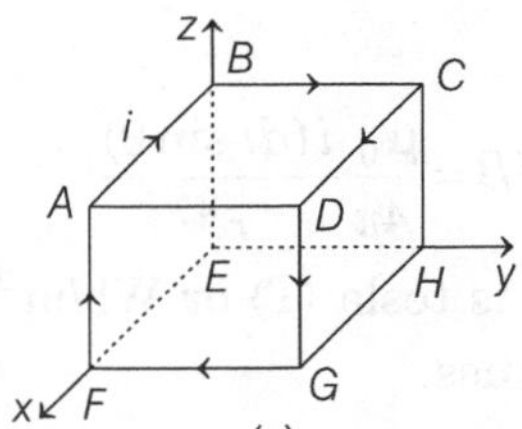

(a)

For example, in Fig. (a), six sides of a cube of length l carry a current i in the directions shown. By assuming two equal and opposite currents in wire AD, two loops in two different planes (xy and yz) are completed.

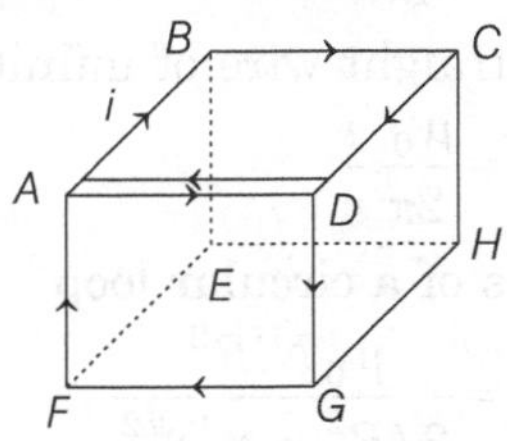

(b)

$$\mathbf{M}_{ABCDA} = -il^2\,\hat{\mathbf{k}}$$

$$\mathbf{M}_{ADGFA} = -il^2\,\hat{\mathbf{i}}$$

$$\therefore \quad \mathbf{M}_{\text{net}} = -il^2\,(\hat{\mathbf{i}} + \hat{\mathbf{k}})$$

- $\dfrac{M}{L} = \dfrac{q}{2m}$ for any system of charges in rotation.

Magnetic Dipole in Uniform Magnetic Field

- $\mathbf{F} = 0$
- $\tau = \mathbf{M} \times \mathbf{B}$ or $\tau = MB\sin\theta$
- $U = -\mathbf{M}\cdot\mathbf{B} = -MB\cos\theta$
- $W_{\theta_1 \to \theta_2} = U_{\theta_2} - U_{\theta_1} = MB(\cos\theta_1 - \cos\theta_2)$
- $\theta = 0°$ is stable equilibrium position of the loop. In this condition,
  $F = 0, \tau = 0$ and $U = -MB =$ minimum
- $\theta = 180°$ is unstable equilibrium position of the loop. Under this condition,
  $F = 0, \tau = 0$ and $U = +MB =$ maximum
- Every current carrying loop is like a bar magnet (or a magnetic dipole).
- Magnetic field due to a bar magnet at a point lying on its axis is
  $$B = \frac{\mu_0}{4\pi}\frac{2M}{r^3} \quad (\text{Along } \mathbf{M})$$
  where, $r >>$ size of bar magnet
- Magnetic field due to a bar magnet at a point on equatorial line is
  $$B = \frac{\mu_0}{4\pi}\frac{M}{r^3} \quad (\text{Opposite to } \mathbf{M})$$
  Again, $r >>$ size of bar magnet

Biot-Savart Law

- $$d\mathbf{B} = \frac{\mu_0}{4\pi}\frac{i(d\mathbf{l} \times \mathbf{r})}{r^3} \quad \text{or} \quad dB = \frac{\mu_0}{4\pi}\frac{i(dl\sin\theta)}{r^2}$$
- SI unit of magnetic field is tesla (T) or Wb/m^2.
  $1\text{T} = 1\ \text{Wb/m}^2 = 10^4$ gauss.

Applications of Biot-Savart Law

- Magnetic field due to a straight wire of finite length,
  $$B = \frac{\mu_0}{4\pi}\frac{i}{r}(\sin\alpha \pm \sin\beta)$$
- Magnetic field due to a straight wire of infinite length,
  $$B = \frac{\mu_0}{2\pi}\frac{i}{r}$$
- Magnetic field on the axis of a circular loop
  $$B = \frac{\mu_0 NiR^2}{2(R^2 + x^2)^{3/2}}$$
- Magnetic field at the centre of a circular loop
  $$B = \frac{\mu_0 Ni}{2R}$$
- Magnetic field at centre due to arc of circle
  $$B = \left(\frac{\theta}{2\pi}\right)\left(\frac{\mu_0 Ni}{2R}\right)$$

- Magnetic field on the axis of a solenoid,

$$B = \frac{\mu_0 ni}{2}(\cos\theta_1 - \cos\theta_2)$$

- Magnetic field at centre (on the axis) of a long solenoid,

$$B = \mu_0 ni$$

- Magnetic field at ends of a long solenoid,

$$B = \frac{\mu_0 ni}{2}$$

Ampere's Circuital Law

$$\oint \mathbf{B} \cdot d\mathbf{l} = \mu_0 (i_{net})$$

Earth's Magnetism

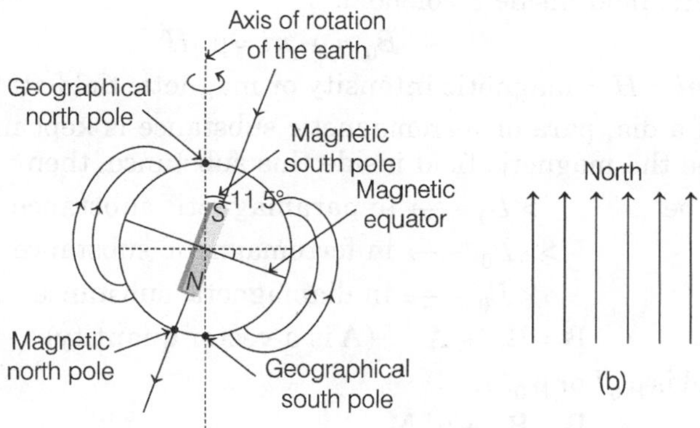

(a) (b)

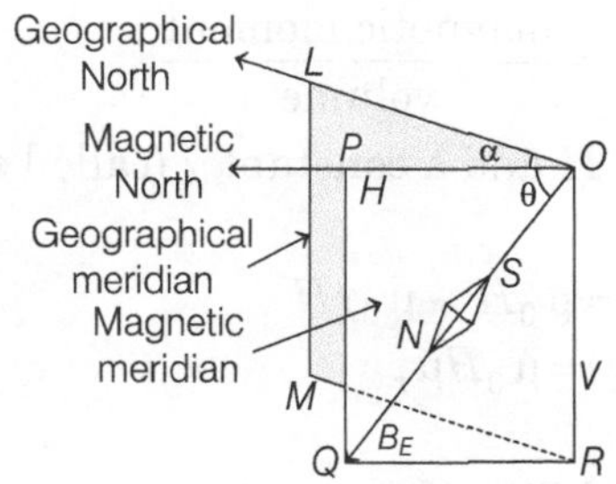

(c)

- The value of magnetic field on the surface of earth is of the order of 10^{-5} T.
- The axis of earth makes an angle of approximately 11.5° with the earth's rotational axis.
- At any point, the vertical plane passing through the line joining the geographical north and south pole is called the geographical meridian.
- At any point, a vertical plane in the direction of earth's magnetic field is called magnetic meridian.
- At any place, the acute angle between the magnetic meridian and geographical meridian is called angle of declination α.

- The angle of dip(θ) at any place is the angle between the direction of earth's magnetic field and the horizontal.

  Magnetic dip will point downward in northern hemisphere (positive dip) and upward in southern hemisphere (negative dip). The range of dip angle is from $-90°$ (at the south magnetic pole) to $+90°$ (at the north magnetic pole).
- $H = B_e \cos\theta$, $V = B_e \sin\theta$, $B_e = \sqrt{H^2 + V^2}$ and $\theta = \tan^{-1}\left(\frac{V}{H}\right)$

  Here, B_e = total earth's magnetic field at some point.

  H = horizontal component of earth's magnetic field at that point

  and V = vertical component of earth's magnetic field at that point.

Magnetic Substances

- Magnetic field inside a solenoid,

$$B_0 = \mu_0 ni = \mu_0 H$$

  Here, $ni = H$ = magnetic intensity or magnetic field strength (A/m)
- Now if a dia, para or ferromagnetic substance is kept inside the solenoid and let B be the magnetic field inside this substance, then

  B may be $> B_0 \longrightarrow$ in paramagnetic substance

  $>> B_0 \longrightarrow$ in ferromagnetic substance

  $< B_0 \longrightarrow$ in diamagnetic substance

  Thus, $\mathbf{B} = \mathbf{B}_0 + \mathbf{A}$ ($\mathbf{A}$ is a vector quantity)

  Here, A is $\mu_0 I$ or $\mu_0 M$.

  or $\mathbf{B} = \mathbf{B}_0 + \mu_0 \mathbf{M}$

  where, I or M is called intensity of magnetisation.

$$I \text{ or } M = \frac{\text{magnetic moment}}{\text{volume}}$$

  I or M depends upon H and a constant χ (called susceptibility).

  So, I or $M = \chi H$
- Thus, $B = B_0 + \mu_0 M = \mu_0 H + \mu_0 \chi H$

  $= \mu_0 H (1 + \chi) = \mu_0 H \mu_r$

  or $B = \mu H$
- $\mu_0 \mu_r = \mu$ = permeability of that substance or

  $\mu_r = \frac{\mu}{\mu_0}$ = relative permeability of that substance
- $B = \mu H$ and $B_0 = \mu_0 H$

  Thus, $\frac{B}{B_0} = \frac{\mu}{\mu_0} = \mu_r = (1 + \chi)$

  or $\chi = \mu_r - 1$
- For diamagnetic substance, $B < B_0$

  $\therefore$ $\mu_r < 1$ or χ is slightly negative.

For paramagnetic substance, $B > B_0$

$\therefore \quad \mu_r > 1$ or χ is slightly positive.

and for ferromagnetic substance, $B >> B_0$

$\mu_r >>$ or χ is highly positive.

- **Curie's law** Magnetic susceptibility of a paramagnetic substance

$$\chi_m \propto \frac{1}{T}$$

Difference between Dia, Para and Ferromagnetic Substances

Properties	Ferromagnetic materials	Paramagnetic materials	Diamagnetic materials
State	They are solid.	They can be solid, liquid or gas.	They can be solid, liquid or gas.
Effect of magnet	Strongly attracted by a magnet.	Weakly attracted by a magnet.	Weakly repelled by a magnet.
Behaviour under non-uniform field	tend to move from low to high magnetic field region.	tend to move from low to high magnetic field region.	tend to move from high to low magnetic field region.
Behaviour under external field	They preserve the magnetic properties after the external field is removed.	They do not preserve the magnetic properties once the external field is removed.	They do not preserve the magnetic properties once the external field is removed.
Effect of temperature	Above Curie point, it becomes a paramagnetic.	With the rise of temperature, it becomes a diamagnetic.	No effect.
Permeability	Very high.	Slightly greater than unity.	Slightly less than unity.
Susceptibility	Very high and positive.	Slightly greater than unity and positive.	Slightly less than unity and negative.
Examples	Iron, Nickel, Cobalt	Lithium and Tantalum	Copper, Silver and Gold

Hysteresis Loop

- For ferromagnetic materials, hysteresis loop is as shown below.

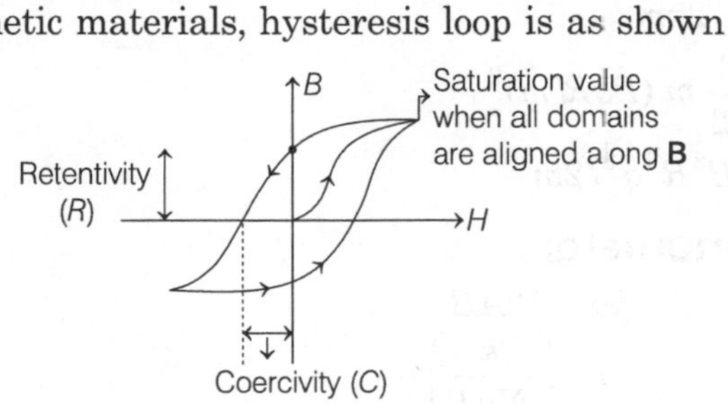

- B is the magnetic field inside the ferromagnetic substance.
- Due to change in direction of H, domains keep on changing their directions. Due to friction, heat is produced.
- Area (A) of the loop is a measure of loss of energy.

- $R_{SI} > R_S \Rightarrow C_{SI} < C_S \Rightarrow A_{SI} < A_S$
  Here, SI is soft iron and S is steel.
- **Permanent magnets** These required high retentivity and coercivity. Hysteresis is immaterial, as this is never put to cyclic changes. So, steel is best suited for it.

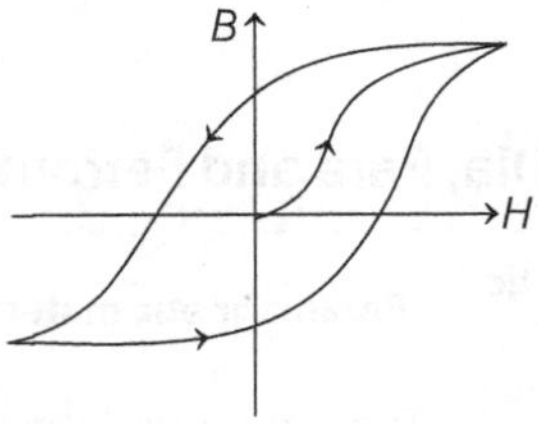

- **Electromagnets** These are temporary magnets. They require high initial permeability and low hysteresis loss. These are used in transformer cores. Soft iron is best suited for it.

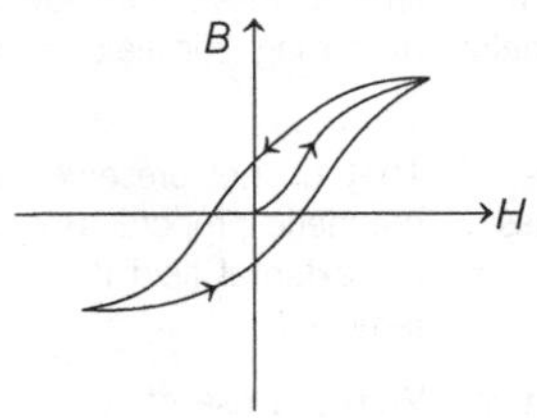

Cyclotron

- $r = \frac{mv}{Bq} \Rightarrow v = \frac{Bqr}{m}$
- $t = \pi r / v = \frac{\pi m}{Bq}$
- $f = \frac{1}{T} = \frac{1}{2t} = \frac{Bq}{2\pi m}$
- $K = \frac{1}{2} mv^2 = \frac{1}{2} m\,(BqR/m)^2$

  $\Rightarrow \quad K = B^2R^2q^2/2m$

Moving Coil Galvanometer

- $$k\phi = NiAB$$
  $$\Rightarrow \quad i = \left(\frac{k}{NAB}\right)\phi$$
- Galvanometer constant $= \frac{k}{NAB}$
- Sensitivity $= \frac{\phi}{i} = \frac{NAB}{k}$

CHAPTER 20

Electromagnetic Induction and Alternating Current

Magnetic Flux

- Magnetic flux $\phi = \int \mathbf{B} \cdot d\mathbf{S}$
- SI unit of magnetic flux is weber.
- If $\mathbf{B}$ is uniform, then

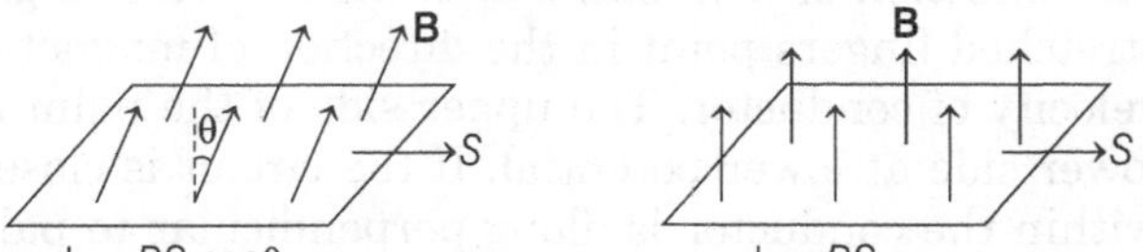

$\phi = BS\cos\theta$ $\qquad$ $\phi = BS$

Faraday's Laws

- Induced emf, $e = -\dfrac{d\phi}{dt}$
- Induced current, $i = e/R = \dfrac{(-d\phi/dt)}{R}$
- Induced flow of charge, $\Delta q = (i\Delta t) = -\dfrac{\Delta\phi}{R}$
- According to Lenz law, any induced event always opposes the cause, due to which it is happening.
- As we have seen, induced emf is produced only when there is a change in magnetic flux passing through a loop. The flux passing through the loop is given by $\phi = BS\cos\theta$.

  This flux can be changed in several ways

  (i) The magnitude of $\mathbf{B}$ can change with time. Thus, $B = B(t)$.

(ii) The current which is producing the magnetic field can change with time. Hence, $i = i(t)$.

(iii) The area of the loop inside the magnetic field can change with time. This can be done by pulling a loop inside (or outside) a magnetic field.

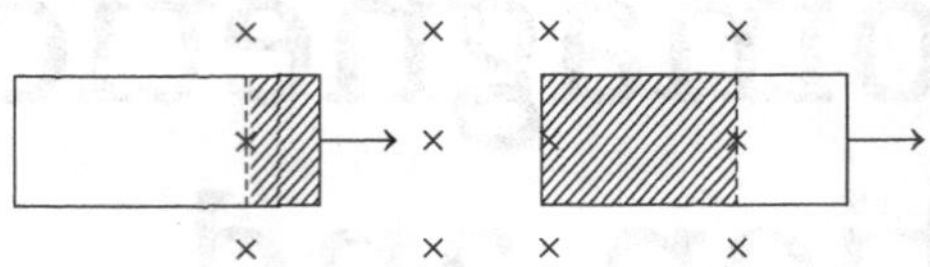

(iv) The angle θ between **B** and the normal to the loop (or **S**) can change with time.

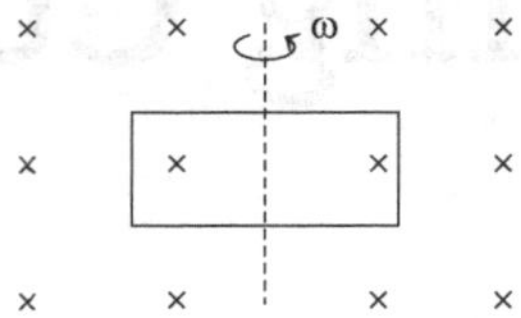

This can be done by rotating a loop in a magnetic field.

Motional EMF

- On a straight conducting wire $e = Bvl$
- On a rotating conducting wire about one end,

$$e = \frac{B\omega l^2}{2}$$

- The direction of motional emf or current can be given by right hand rule. The stretched fingers point in the direction of magnetic field. Thumb is along the velocity of conductor. The upper side of the palm is at higher potential and lower side at lower potential. If the circuit is closed, the induced current within the conductor is along perpendicular to palm upwards.

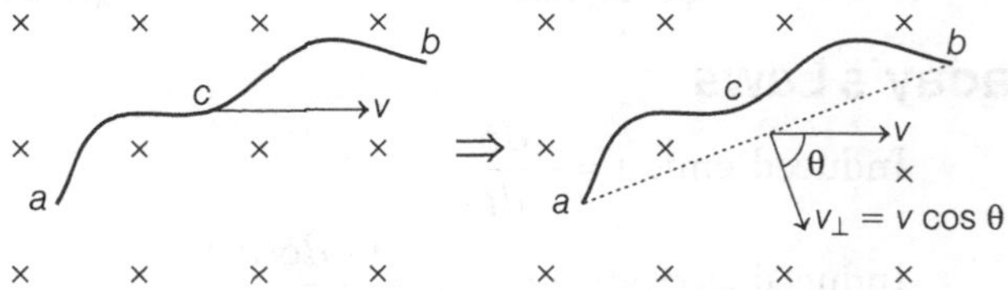

In the above figure, from right hand rule we can see that b is at higher potential and a at lower potential.

$$V_{ba} = V_b - V_a = (ab)(v\cos\theta)(B)$$

Self and Mutual Induction

- Coefficient of self induction, $L = \dfrac{N\phi}{i}$ or $\left(\dfrac{-e}{di/dt}\right)$

- Coefficient of self induction of a circular coil,
$$L \propto N^2$$
- Potential difference across an inductor,
$$\text{PD} = -L\left(\frac{di}{dt}\right)$$
- Energy stored in the magnetic field of inductor,
$$U = \frac{1}{2}Li^2$$
- Coefficient of mutual inductance,
$$M = \frac{N_2\,\phi_2}{i_1} \quad \text{or} \quad \left(\frac{-e_2}{di_1/dt}\right)$$
- Coefficient of mutual inductance of two closely wound circular coils,
$$M \propto N_1 N_2$$
- Inductors in series, $L = L_1 + L_2$
- Inductors in parallel, $\frac{1}{L} = \frac{1}{L_1} + \frac{1}{L_2}$ or $L = \frac{L_1 L_2}{L_1 + L_2}$
- Energy density or energy per unit volume in magnetic field,
$$u = \frac{1}{2}\frac{B^2}{\mu_0}$$
- **Kirchhoff's potential law with an inductor**
In Kirchhoff's potential equation when we jump an inductor in the direction of current, we encounter a voltage drop equal to $L\,di/dt$, where di/dt is to be substituted with sign.

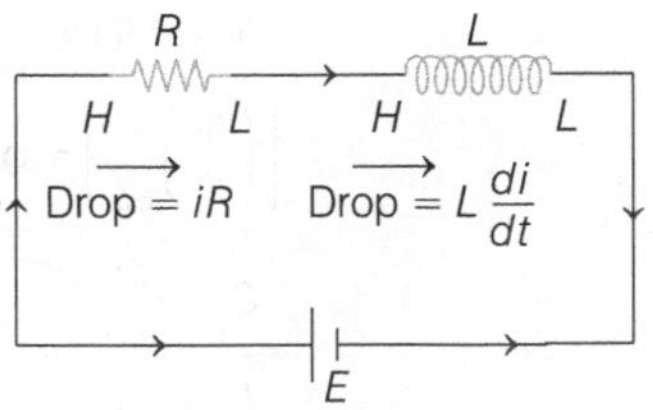

For example, in the loop shown in figure, Kirchhoff's second law gives the equation.
$$E - iR - L\frac{di}{dt} = 0$$
- Current growth in an L-R circuit,
$$i = i_0\,(1 - e^{-t/\tau_L})$$
- Current decay in L-R circuit, $i = i_0\, e^{-t/\tau_L}$
- τ_L = time constant $= \frac{L}{R}$
- Unit of τ_L or τ_C is second.

Oscillations in *L-C* circuit

In L - C circuit, charge, current and rate of change of current oscillate simple harmonically. These oscillations are similar to oscillations of spring-block system in the chapter of SHM. Table below shows a comparison of oscillations of a mass-spring system and L - C circuit.

A comparison of oscillations of a mass-spring system and *L-C* circuit

Mass-spring system	Inductor-capacitor circuit
Displacement (x)	Charge (q)
Velocity (v)	Current (i)
Acceleration (a)	Rate of change of current $\left(\frac{di}{dt}\right)$
$\frac{d^2x}{dt^2} - \omega^2 x$, where $\omega = \sqrt{\frac{k}{m}}$	$\frac{d^2q}{dt^2} = -\omega^2 q$, where $\omega = \frac{1}{\sqrt{LC}}$
$x = A\sin(\omega t \pm \phi)$ or $x = A\cos(\omega t \pm \phi)$	$q = q_0 \sin(\omega t \pm \phi)$ or $q = q_0 \cos(\omega t \pm \phi)$
$v = \frac{dx}{dt} = \omega\sqrt{A^2 - x^2}$	$i = \frac{dq}{dt} = \omega\sqrt{q_0^2 - q^2}$
$a = \frac{dv}{dt} = -\omega^2 x$	Rate of change of current $= \frac{di}{dt} = -\omega^2 q$
Kinetic energy $= \frac{1}{2}mv^2$	Magnetic energy $= \frac{1}{2}Li^2$
Potential energy $= \frac{1}{2}kx^2$	Potential energy $= \frac{1}{2}\frac{q^2}{C}$
$\frac{1}{2}mv^2 + \frac{1}{2}kx^2 = \text{constant} = \frac{1}{2}kA^2 = \frac{1}{2}mv_{max}^2$	$\frac{1}{2}Li^2 + \frac{1}{2}\frac{q^2}{C} = \text{constant} = \frac{1}{2}\frac{q_0^2}{C} = \frac{1}{2}Li_{max}^2$
$\lvert v_{max} \rvert = A\omega$	$i_{max} = q_0\omega$
$\lvert a_{max} \rvert = \omega^2 A$	$\left\lvert \left(\frac{di}{dt}\right)_{max} \right\rvert = \omega^2 q_0$
$\frac{1}{k}$	C
m	L

Induced Electric Field

- The line integral of **E** around a closed path is not zero. This line integral is given by

$$\oint \mathbf{E} \cdot d\mathbf{l} = -\frac{d\phi_B}{dt}$$

This electric field is produced by change in magnetic field.

- Being a non-conservative field, the concept of potential has no meaning for such a field.
- This field is different from the electrostatic field produced by stationary charges (which is conservative in nature).
- The relation $\mathbf{F} = q\mathbf{E}$ is valid for this field.

Eddy Currents

- When a changing magnetic flux is applied to a piece of conducting material, circulating currents called eddy currents are induced in the material. These eddy currents often have large magnitudes and heat up the conductor.
- When a metal plate is allowed to swing through a strong magnetic field, then in entering or leaving the field, the eddy currents are set up in the plate which opposes the motion as shown in figure.

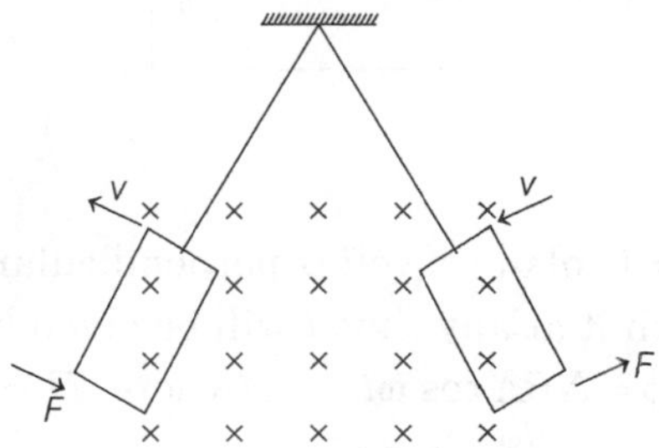

- The kinetic energy dissipates in the form of heat. The slowing down of the plate is called the **electromagnetic damping.**
- The electromagnetic damping is used to damp the oscillations of a galvanometer coil or chemical balance and in braking electric trains. Otherwise, the eddy currents are often undesirable.
- To reduce the eddy currents some slots are cut into moving metallic parts of machinery. These slots intercept the conducting paths and decreases the magnitudes of the induced currents.

Back EMF of Motors

- An electric motor converts electrical energy into mechanical energy and is based on the fact that a current carrying coil in a uniform magnetic field experiences a torque.
- As the coil rotates in the magnetic field, the flux linked with the rotating coil will change and hence, an emf called back emf is produced in the coil.
- When the motor is first turned ON, the coil is at rest and so there is no back emf. The 'start up' current can be quite large. To reduce 'start up' current a resistance called 'starter' is put in series with the motor for a short period when the motor is started.
- As the rotation rate increases the back emf increases and hence, the current reduces.

Electric Generator or Dynamo

- A dynamo converts mechanical energy (rotational kinetic energy) into electrical energy. It consists of a coil rotating in a magnetic field.

- Due to rotation of the coil, magnetic flux linked with it changes, so an emf is induced in the coil.

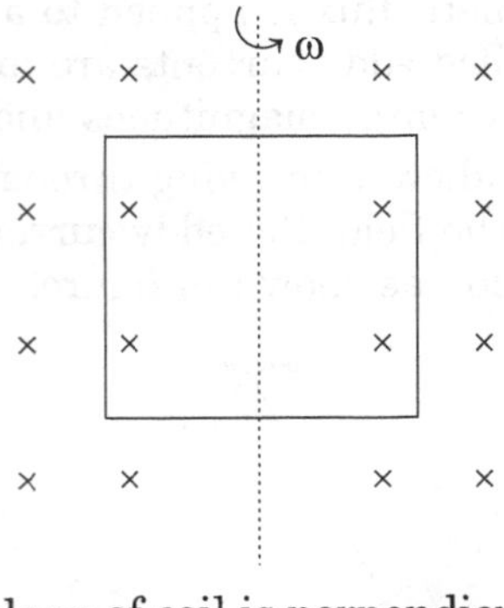

Suppose at time $t = 0$, plane of coil is perpendicular to the magnetic field. The flux linked with it at any time t will be given by

$$\phi = NBA \cos \omega t \qquad \text{(where, } N = \text{number of turns in the coil)}$$

$$\therefore \qquad e = -\frac{d\phi}{dt} = NBA\omega \sin \omega t \text{ or } e = e_0 \sin \omega t$$

where, $e_0 = NBA\,\omega$

Alternating Current

- Frequency of AC in India is 50 Hz.
- The AC is converted into DC with the help of rectifier while DC is converted into AC with the help of inverter.
- An AC cannot produce electroplating or electrolysis.
- The AC is measured by hot wire ammeter.
- An AC can be stepped up or down with the help of a transformer.
- An AC can be transmitted over long distances without much power loss.
- An AC can be regulated by using choke coil without any significant waste of energy.
- In an AC (sinusoidal) current or voltage can have following four values :
  (i) instantaneous value
  (ii) peak value (i_0 or V_0)
  (iii) irms value (i_{rms} or V_{rms})
  (iv) average value; (In full cycle, average value is zero while in half cycle it is non-zero.)
- $i_{rms} = \dfrac{i_0}{\sqrt{2}}, V_{rms} = \dfrac{V_0}{\sqrt{2}}$
- $\langle i \rangle_{\text{positive half cycle}} = \dfrac{2}{\pi} i_0 \approx 0.636\, i_0$

  Similarly, $\langle V \rangle_{\text{positive half cycle}} = \dfrac{2}{\pi} V_0 \approx 0.636\, V_0$

Note *In sinusoidal AC, even the average value in half cycle can also be zero. It depends on the time interval over which half, average value is desired.*

Series *L-C-R* AC circuit

- Capacitive reactance, $X_C = \frac{1}{\omega C}$
- Inductive reactance, $X_L = \omega L$
- Impedance, $Z = \sqrt{R^2 + (X_C - X_L)^2}$
- If $X_C > X_L$, current leads and if $X_L > X_C$, voltage leads by an angle ϕ given by

$$\cos\phi = \frac{R}{Z}$$

or
$$\tan\phi = \frac{X_C \sim X_L}{R}$$

- Instantaneous power = instantaneous current × instantaneous voltage
- Average power = $V_{\text{rms}}\, i_{\text{rms}} \cos\phi$, where $\cos\phi = \frac{R}{Z}$ = power factor.
- $i_0 = \frac{V_0}{Z}$ or $i_{\text{rms}} = \frac{V_{\text{rms}}}{Z}$
- $(V_C)_{\text{rms}} = (i_{\text{rms}})\, X_C$, $(V_L)_{\text{rms}} = (i_{\text{rms}})\, X_L$ and $(V_R)_{\text{rms}} = (i_{\text{rms}})\, R$
- $V = \sqrt{V_R^2 + (V_C - V_L)^2}$

  Here, V is the rms value of applied voltage,
  V_R is the rms value of voltage across resistance,
  V_C across capacitor and V_L across inductor, etc.
- At $\omega = \omega_r = \frac{1}{\sqrt{LC}}$, $X_C = X_L$ and Z has the minimum value equal to R. Power factor in this case is 1.
- At $\omega > \omega_r$, $X_L > X_C$, voltage leads the current function and circuit is inductive.
- At $\omega < \omega_r$, $X_C > X_L$, current leads the voltage function and circuit is capacitive.
- **Quality factor** $Q = \frac{\omega_r L}{R}$

  It is a measure of sharpness of resonance. If value of Q is large, sharpness is more and it is more selective.

Transformer

- It is a device which is either used to increase or decrease the voltage in AC circuits through mutual induction. A transformer consists of two coils wound on the same core.
- The coil connected to input is called primary while the other connected to output is called secondary coil. An alternating current passing through the primary creates a continuously changing flux through the core.

This changing flux induces an alternating emf in the secondary.

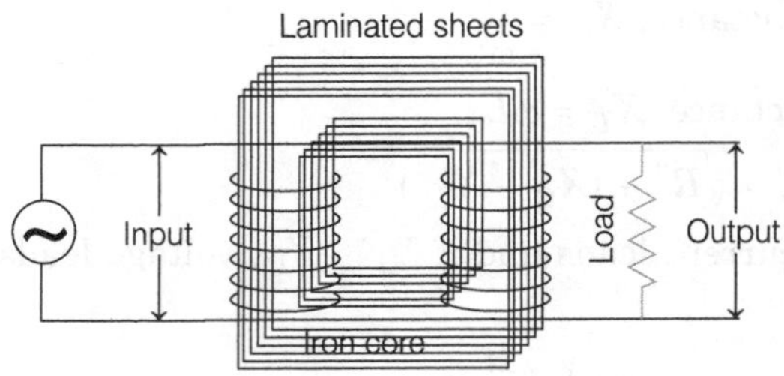

- As magnetic lines of force are closed curves, the flux per turn of primary must be equal to flux per turn of the secondary. Therefore,

$$\frac{\phi_P}{N_P} = \frac{\phi_S}{N_S}$$

or $$\frac{1}{N_P} \cdot \frac{d\phi_P}{dt} = \frac{1}{N_S} \cdot \frac{d\phi_S}{dt} \qquad \left(\text{As } e \propto \frac{d\phi}{dt}\right)$$

$\therefore$ $$\frac{e_S}{e_P} = \frac{N_S}{N_P}$$

- In an ideal transformer, there is no loss of power.

Hence, ei = constant

$\Rightarrow$ $$\frac{e_S}{e_P} = \frac{N_S}{N_P} = \frac{i_P}{i_S}$$

- Regarding a transformer, given below are few important points
  (i) In step-up transformer, $N_S > N_P$. It increases voltage and reduces current.
  (ii) In step-down transformer, $N_P > N_S$. It increases current and reduces voltage.
  (iii) It works only on AC.
  (iv) A transformer cannot increase (or decrease) voltage and current simultaneously. As, ei = constant
  (v) Some power is always lost due to eddy currents, hysteresis, etc.

CHAPTER 21

Modern Physics

Bohr's Theory

- Bohr's theory is applicable for hydrogen and hydrogen like atoms/ions. For such type of atoms/ions, number of electrons is one. Although atomic numbers may be different.

  Example For ${}_1H^1$, atomic number $Z = 1$, for He^+, atomic number $Z = 2$ and for Li^{+2}, atomic number $Z = 3$.

  But for all three, number of electrons is one.
- In nth orbit,

$$\frac{mv^2}{r} = \frac{1}{4\pi\varepsilon_0} \cdot \frac{(e)(Ze)}{r^2}$$

and

$$L_n = mvr = \frac{nh}{2\pi}$$

After solving above these two equations, we will get following results

(i) $r \propto \frac{n^2}{Z}$ and $r \propto \frac{1}{m}$

(ii) $v \propto \frac{Z}{n}$ and $v \propto m^0$

(iii) $E \propto \frac{Z^2}{n^2}$ and $E \propto m$

(iv) $r_1^H = 0.529$ Å

(v) $v_1^H = 2.2 \times 10^6$ m/s $= \frac{c}{137}$

(vi) $E_1^H = -13.6$ eV

(vii) $K = |E|$ and $U = 2E$

- Energy of a photon $E = \frac{hc}{\lambda}$. After substituting values of h and c, we get

$$E \text{ (in eV)} = \frac{12375}{\lambda \text{ (in Å)}}$$

- **Hydrogen spectrum**

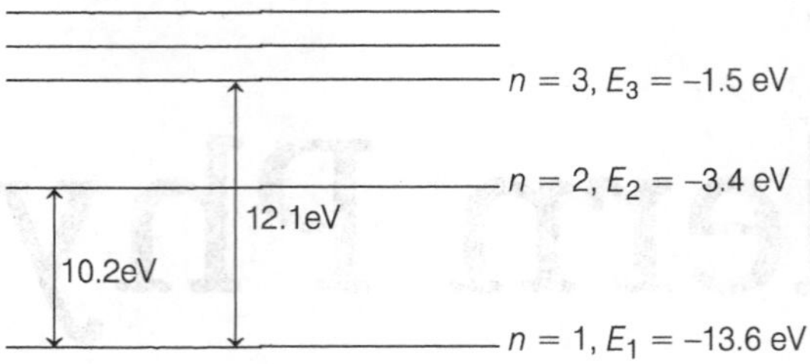

- In first orbit of hydrogen atom, $E = -13.6$ eV, $K = 13.6$ eV and $U = -27.2$ eV
  Similarly, in second orbit, $E = -3.4$ eV, $K = 3.4$ eV and $U = -6.8$ eV
- $n = 1$ is ground state, $n = 2$ is first excited state and $n = 3$ is second excited state.
- If an electron jumps from n_1 state to n_2 $(< n_1)$ state, then wavelength of photon emitted in this process will be given by

$$\frac{hc}{\lambda} = E_{n_1} - E_{n_2}$$

or $$\lambda(\text{in Å}) = \frac{12375}{(E_{n_1} - E_{n_2}) \text{ in eV}}$$

Here, $$E_{n_1} = \frac{-13.6 \times Z^2}{n_1^2} \text{ eV}$$

and $$E_{n_2} = \frac{-13.6 \times Z^2}{n_2^2} \text{ eV}$$

Note $\frac{1}{\lambda} = RZ^2\left(\frac{1}{n_2^2} - \frac{1}{n_1^2}\right)$

- $Rhc = 1$ Rydberg $= 13.6$ eV
- $R =$ Rydberg constant $= 1.09 \times 10^7 \text{ m}^{-1}$
- As n increases
  (i) Angular momentum, time period, potential energy and total energy will increase.
  (ii) Speed, kinetic energy and frequency will decrease.
- Total number of emission lines from nth state to ground state are $\frac{n(n-1)}{2}$.
- Whenever the force obeys inverse square law $\left(F \propto \frac{1}{r^2}\right)$ and potential energy is inversely proportional to r, $\left(U \propto \frac{1}{r}\right)$, then kinetic energy ($K$), potential energy ($U$) and total energy ($E$) have the following relationships

$$K = \frac{|U|}{2} \quad \text{and} \quad E = -K = \frac{U}{2}$$

Matter Wave or de-Broglie Wave

- Every matter particle having some linear momentum is associated with a wave called matter wave or de-Broglie wave.
- Wavelength of this wave is given by

$$\lambda = \frac{h}{p} = \frac{h}{mv} = \frac{h}{\sqrt{2Km}} = \frac{h}{\sqrt{2qVm}}$$

- For an electron, $\lambda \text{ (in Å)} = \sqrt{\frac{150}{V \text{ (in volt)}}} = \sqrt{\frac{150}{\text{KE(in eV)}}}$
- In hydrogen and hydrogen like atoms,
  Circumference of nth orbit $= n$ (wavelength of single electron in that orbit)
  It can be understand as, according to Bohr's assumption,

$$L_n = n\left(\frac{h}{2\pi}\right)$$

or

$$mvr = n\left(\frac{h}{2\pi}\right)$$

$\therefore$

$$(2\pi r) = n\left(\frac{h}{mv}\right)$$

or circumference $= n$ (de-Broglie wavelength of electron)

Electromagnetic Waves

- EM waves have dual nature, particle as well as a wave.
- Interference, diffraction or polarisation can be explained by wave nature of EM waves.
  Photoelectric effect, Compton effect, etc. can be explained by particle nature of EM waves.
- **E**, **B** and **c** are mutually perpendicular.
  $\therefore \quad \mathbf{E} = \mathbf{B} \times \mathbf{c}$
  and E and B oscillate in same phase.
  For example, $B_x = B_0 \sin(\omega t - ky)$, then $E_z = E_0 \sin(\omega t - ky)$
- $\omega = 2\pi f,\ T = \frac{1}{f} = \frac{2\pi}{\omega}$
- $c = \frac{E_0}{B_0}$
- **Properties of a photon**
  (i) $E = hf = \frac{hc}{\lambda}$
  (ii) $E \text{ (in eV)} = \frac{12375}{\lambda \text{ (in Å)}}$
  (iii) Momentum of photon, $p = \frac{E}{c} = \frac{h}{\lambda}$

- In moving from left to right in this spectrum, energy and frequency of wave decrease but wavelength increases.

  Electromagnetic spectrum constitutes; γ-rays, X-rays, UV rays, visible light, infrared rays, microwaves, radio waves.

- **Radiation pressure**

$$p = \frac{I}{C} \quad \text{(For totally absorbed surface)}$$

and

$$p = \frac{2I}{C} \quad \text{(For totally reflecting surface)}$$

X-Rays

- These are electromagnetic waves of high energy, high frequency and low wavelength.
- Wavelength of X-rays lies in the range of 0.1 Å to 100 Å.
- Cut off wavelength of X-ray spectrum is given by

$$\lambda_{min} \text{ (in Å)} = \frac{12375}{V \text{ (in volt)}}$$

- Moseley's law for X-ray spectrum is

$$\sqrt{\nu} \propto (Z - b)$$

  Here, the constant b depends upon the series

$$b = 1, \text{ for } K\text{-series}$$

and

$$b = 7.4, \text{ for } L\text{-series}$$

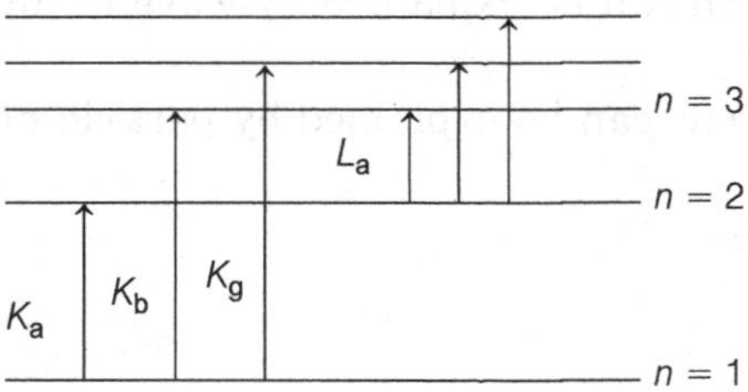

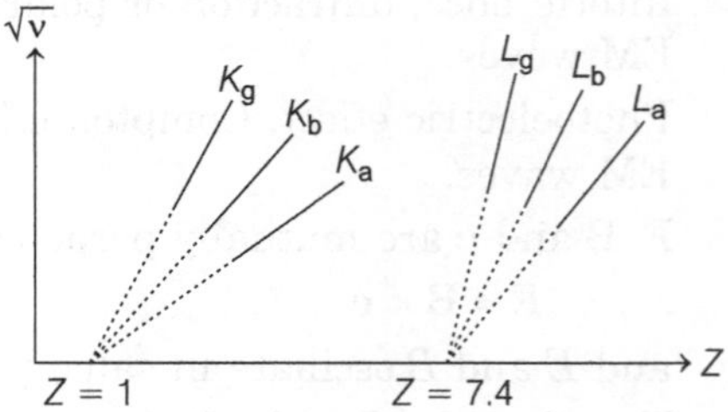

- $$\frac{1}{\lambda} = (Z - b)^2 R\left(\frac{1}{n_1^2} - \frac{1}{n_2^2}\right)$$

  Here, $n_1 < n_2$

- **Continuous X-Ray spectrum**

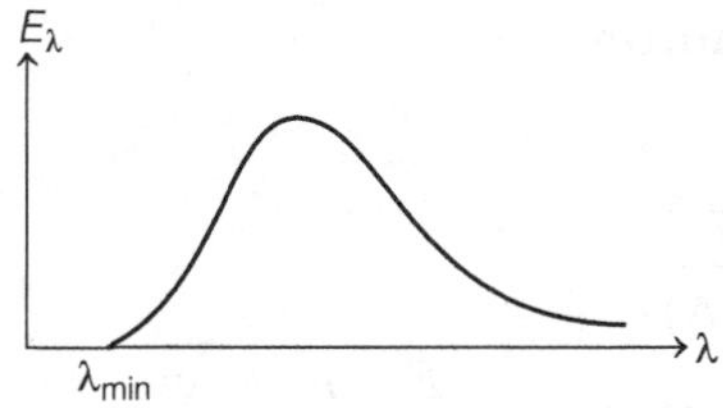

Here, $\lambda_{min} \text{ (in Å)} = \frac{12375}{V \text{ (in volt)}}$ = cut-off wavelength.

- This is actual spectrum of X-rays including continuous and characteristic X-rays.

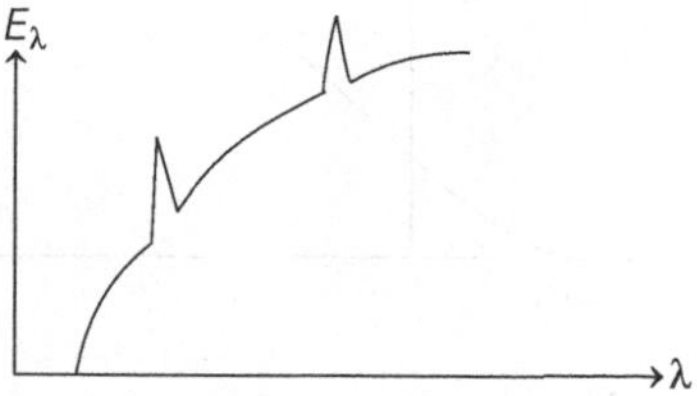

Photoelectric Effect

- $K_{max} = \frac{1}{2} m v_{max}^2 = E - W = hf - hf_0 = eV_0 = \frac{hc}{\lambda} - \frac{hc}{\lambda_0}$
- **K_{max} *versus* f graph**

  From the equation $K_{max} = hf - W$, we can see that K_{max} *versus* f graph is a straight line with positive slope h (a universal constant) and negative intercept W (varies metal to metal).

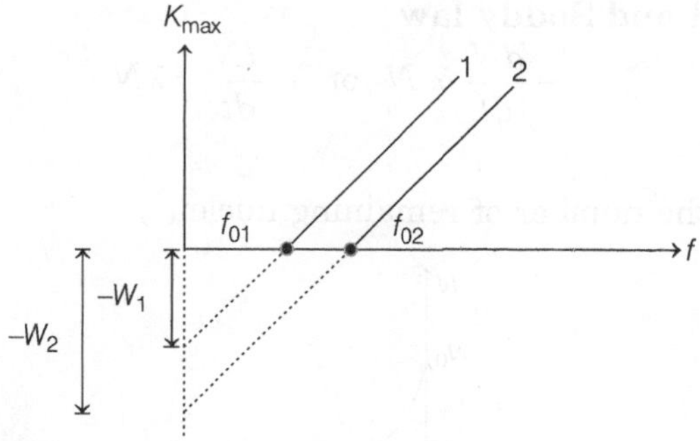

$$\begin{aligned} \text{Slope}_1 &= \text{Slope}_2 = \text{Planck's constant,} \\ f_{01} &= \text{Threshold frequency of metal-1,} \\ f_{02} &= \text{Threshold frequency of metal-2,} \\ W_1 &= \text{Work function of metal-1 and} \\ W_2 &= \text{Work function of metal-2.} \end{aligned}$$

- Threshold wavelength, $\lambda_0 = \frac{hc}{W}$
- Threshold frequency, $f_0 = \frac{W}{h}$
- **Photoelectric current *versus* potential graph**

  Saturation current I_s depends upon the number of photons incident on metallic plate per second and stopping potential V_0 depends upon the frequency of incident photons.

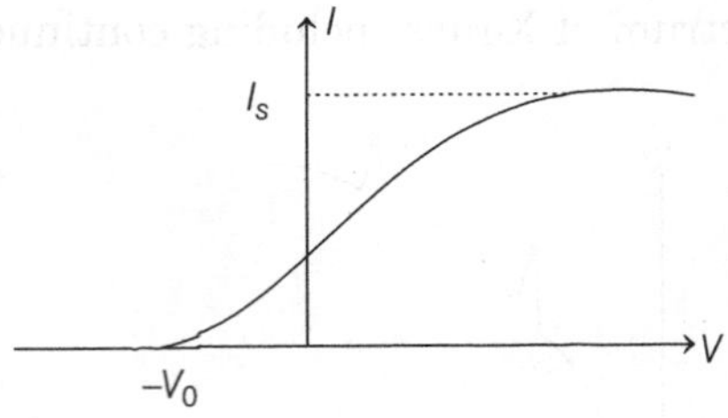

- .For photoemission to take place,

$$E \geq W, \lambda \leq \lambda_0 \text{ and } f \geq f_0$$

- Kinetic energy of photoelectrons varies between 0 and K_{max}.

Radioactivity

- α-emission, ${}_Z X^A \xrightarrow{\alpha\text{-emission}} {}_{Z-2}Y^{A-4}$
- β-emission, $n \longrightarrow p + e^- + \bar{\nu}$

$\Rightarrow \quad {}_Z X^A \xrightarrow{\beta\text{-emission}} {}_{Z+1}Y^A$

- **Rutherford and Soddy law**

$$-\frac{dN}{dt} \propto N \quad \text{or} \quad -\frac{dN}{dt} = \lambda N$$

- $N = N_0 \, e^{-\lambda t}$

Here, N is the number of remaining nuclei.

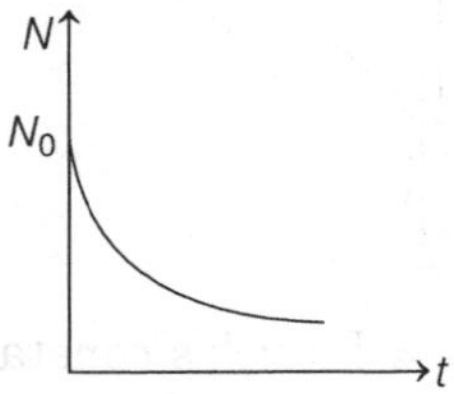

- $N_d = N_0 \, (1 - e^{-\lambda t})$

Here, N_d is the number of decayed nuclei.

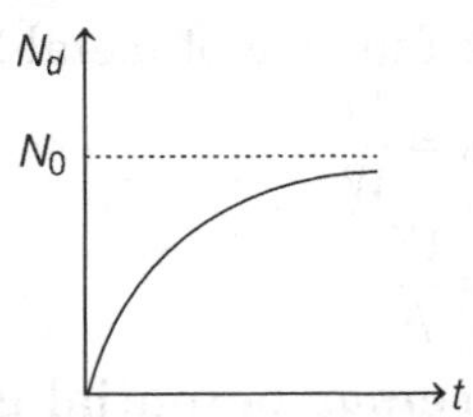

- λ = decay constant

$$\frac{1}{\lambda} = \text{mean life or average life} = t_{av}$$

$$t_{1/2} = (\ln 2)\left(\frac{1}{\lambda}\right) = \frac{0.693}{\lambda}$$

- $t_{av} > t_{1/2}$

$$t_{1/2} = \frac{\ln 2}{\lambda} = \frac{0.693}{\lambda} = 0.693\, t_{av}$$

$$t_{av} = \frac{1}{\lambda} = 1.44\, t_{1/2}$$

- Probability of survival of a nucleus up to time t

$$P_s = e^{-\lambda t}$$

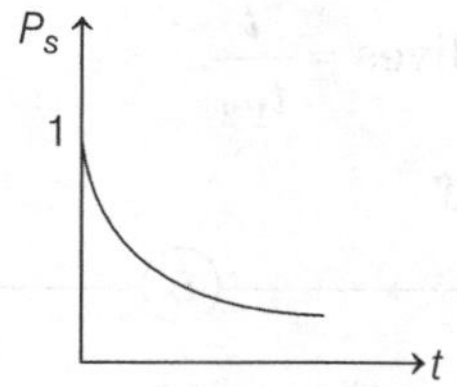

- Probability of decay of a nucleus in time t,

$$P_d = 1 - e^{-\lambda t}$$

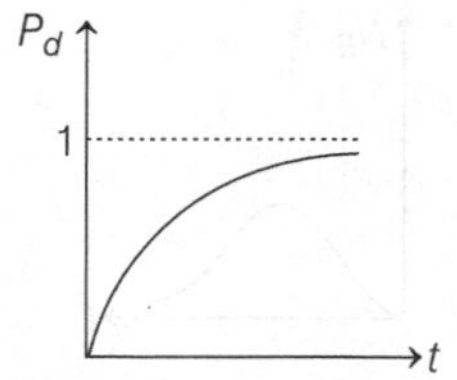

- Activity of radioactive substance,

$$R = -\frac{dN}{dt} = \lambda N$$

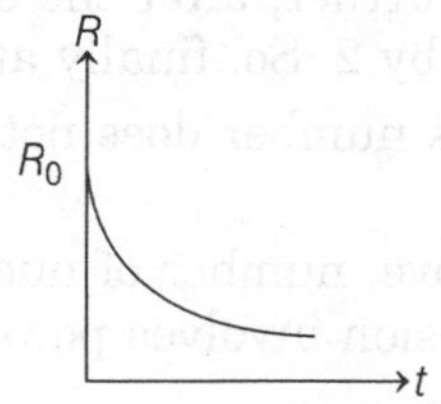

$$= \lambda N_0 e^{-\lambda t} = R_0 e^{-\lambda t}$$

Here, $R_0 = \lambda N_0$ = initial activity.

- Units of activity
  - (i) 1 Curie = 1 Ci = 3.7×10^{10} dps
  - (ii) 1 Rutherford = 1 rd = 10^6 dps
  - (iii) 1 Becquerel = 1 Bq = 1 dps
- If a nucleus decays by two modes, then

$$\lambda = \lambda_1 + \lambda_2 \text{ and } T = \frac{T_1 T_2}{T_1 + T_2}$$

Here, T is the half-life.

- $N_0 \xrightarrow{t_{1/2}} \frac{N_0}{2} \xrightarrow{t_{1/2}} \frac{N_0}{4} \ldots\ldots \left(\frac{1}{2}\right)^n N_0$

  $1 \longrightarrow \frac{1}{2} \longrightarrow \frac{1}{4} \ldots\ldots \left(\frac{1}{2}\right)^n$

  $100\% \longrightarrow 50\% \longrightarrow 25\% \ldots 100\left(\frac{1}{2}\right)^n$

  Here, n = number of half-lives $= \frac{t}{t_{1/2}}$.

- **Successive radioactivity**

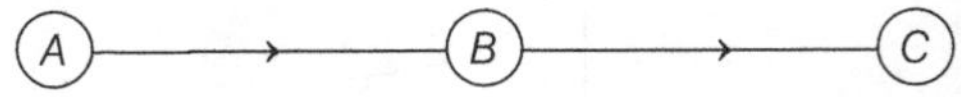

Suppose A and B are radioactive substances and C is stable. Let us further assume that initially there are only nuclei of A. Then, number of nuclei of A, B and C vary with time as shown below.

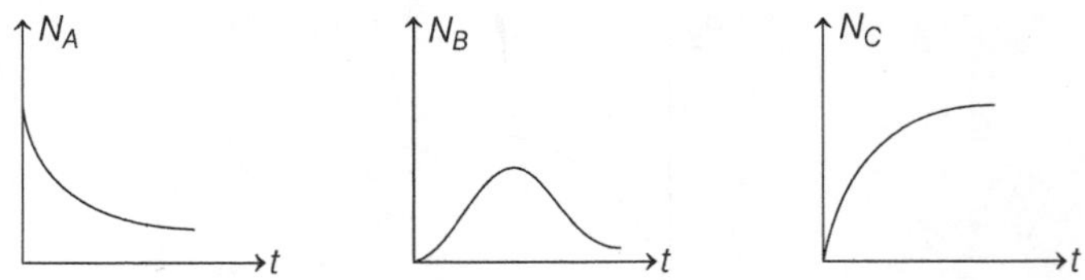

N_B are maximum when $\lambda_A N_A = \lambda_B N_B$.

- After emission of one alpha particle and two beta particles isotopes are produced. This is because after the emission of one alpha particle, atomic number decreases by 2. Further, after the emission of two beta particles, atomic number increases by 2. So, finally atomic number remains unchanged.
- From beta emission, mass number does not change. Therefore, isobars will be produced.
- As we have discussed above, number of nuclei decayed in time t are $N_0 (1 - e^{-\lambda t})$. This expression involves power of e.

  So, to avoid it, we can use

  $$-\Delta N = \lambda N \Delta t$$

  where, $-\Delta N$ are the number of nuclei decayed in time Δt at the instant when total number of nuclei are N. But, this can be applied only when $\Delta t << t_{1/2}$.

  Proof $$-\frac{dN}{dt} = \lambda N$$

  $\Rightarrow$ $$-dN = \lambda N dt$$

  or $$-\Delta N = \lambda N \Delta t$$

- In same interval of time, equal percentage (or fraction) of nuclei are decayed (or left undecayed).

Fusion and Fission

- $E = mc^2$
- $1 \text{ amu} = 931.48 \dfrac{\text{MeV}}{c^2}$
- During formation of a nucleus, some mass is lost. This is called mass defect. Equivalent to that mass, some energy is liberated. This is called binding energy of the nucleus.
- Mass defect, $\Delta m = [Zm_p + (A - Z)\, m_N - m_x]$
  Here m_x is mass of nucleus.
- Binding energy $E_b = (\Delta m)\, c^2$
  If Δm is represented in amu, then E_b can be obtained in MeV by multiplying it with 931.48.
- Binding energy per nucleon
$$\frac{E_b}{A} = \frac{E_b}{\text{Total number of nucleons}}$$
- For stability of a nucleus binding energy per nucleon is more important rather than the total binding energy.
- Binding energy per nucleon is of the order of few MeV (2 MeV-10 MeV).
- In any nuclear process energy is released, if total binding energy of the daughter nuclei is more than the total binding energy of the parent nuclei.
- **Binding energy per nucleon *versus* mass number graph**

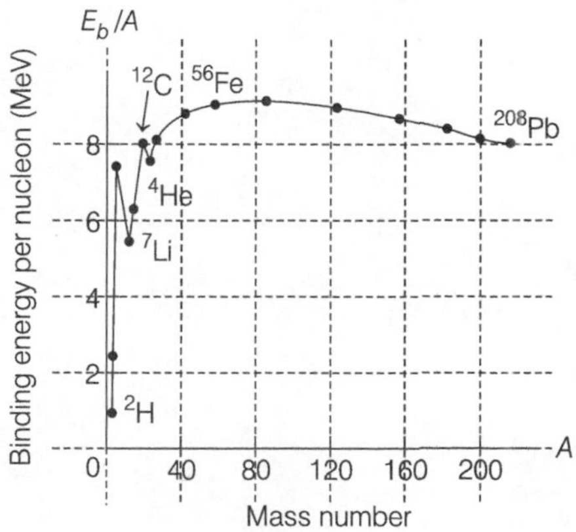

In any nuclear process, if the products are towards the peak of this graph then binding energy per nucleon, hence total binding energy will increase. Therefore, energy will be released.

In fusion reaction, two or more lighter nuclei combine to make a relatively heavier nucleus. This heavy nucleus is towards peak of this graph. Therefore, energy will be released.

In fission reaction, a heavy nucleus breaks into two or more lighter nuclei. These lighter nuclei again lie towards peak of this graph.

Therefore, energy will be released.

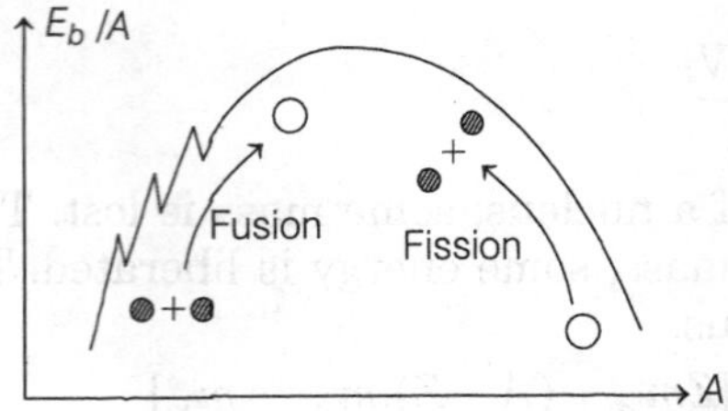

- Normally, fusion reaction is more difficult compared to fission reaction. Because to combine two or more positively charged nuclei is difficult compared to break it.

  Here, $R_0 = 1.3 \text{ fm} = 1.3 \times 10^{-15}$ m

  and A = mass number.

  Thus, $R \propto A^{1/3}$
- The density of any nucleus is independent of A and is of the order of 10^{17} kg m^{-3}.

CHAPTER 22

Semiconductors

Band Theory of Solids

- Energy band theory is used to study the behaviour of solids as conductors (metals), insulators and semiconductors.
- Electrons of each isolated atom have discrete energy levels. When two similar atoms are brought closer, then there is an interaction between the electrons of these two. This interaction causes a splitting of each individual energy level into two slightly different energy levels.
- The atoms in a solid are so close to each other that the energy levels produced after splitting due to interaction between the various atoms will be so close to each other that they appear as continuous. These closely spaced energy levels form an energy band.
- **How these bands form?**

  Consider a silicon (Si) crystal having N atoms.

  (i) The electronic configuration of silicon ($Z = 14$) is

  $$1s^2, 2s^2, 2p^6, 3s^2, 3p^2$$

  (ii) The energy levels of electrons in the silicon atom are shown in figure.

  (iii) The levels $1s$, $2s$, $2p$ and $3s$ are completely filled and the level $3p$ contains only 2 electrons, whereas it can accommodate 6 electrons.

  (iv) In a silicon crystal, there are $14N$ electrons. For each silicon atom, there are two states in energy level $1s$. So there are $2N$ states in energy level $1s$, $2N$ states in energy level $2s$, $6N$ states in energy level $2p$, $2N$ states in energy level $3s$ and $6N$ states in $3p$ energy level of a silicon crystal. In $3p$ energy level only $2N$ states are filled and $4N$ states are empty.

- What happens when interatomic spacing (r) decreases?

  There are four conditions to understand the interatomic spacing which are given in the following table.

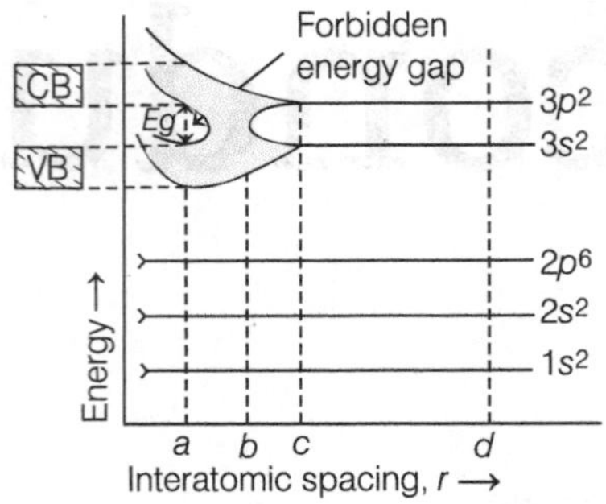

At $r = d >> a$ (where a = crystal lattice spacing), i.e. when interatomic spacing is very large	Each atom in the silicon crystal behaves independently and has discrete energy levels.
At $r = c$	The interaction among valence electrons of N atoms splits $3s$ and $3p$ levels into large number of closely spaced energy levels where the energy of an electron may be slightly less or more than the energy of an electron in an isolated atom. Thus, two bands corresponding to $3s$ and $3p$ states are formed. As the interatomic spacing (r) is further decreased, the energy bands corresponding to $3s$ and $3p$ states spread more and hence energy gap between these bands decreases. The levels $1s^2$, $2s^2$ and $2s^6$ lie in the interior of an atom and hence, cannot be influenced by other atoms.
At $r = b$	The $3s$ and $3p$ bands overlap and the energy gap between them disappears. In this case, all $8N$ levels ($2N$ corresponding to $3s$ energy level and $6N$ corresponding to $3p$ energy level) are now continuously distributed. Out of these $8N$ levels, $4N$ levels are filled and $4N$ are empty.
At $r = a$, i.e. the equilibrium separation.	The filled and unfilled energy levels are separated by an energy gap called forbidden energy gap. The energy gap is denoted by Eg. The lower filled energy band is called valence band and the upper unfilled energy band is called conduction band.

Note *No electron of a solid can stay in a forbidden energy gap as there is no allowed energy state in this region. The width of the forbidden energy gap is a measure of the bondage of valence electrons to the atom.*

- A solid can behave as a conductor, insulator or a semiconductor depending on the width of the forbidden energy band.

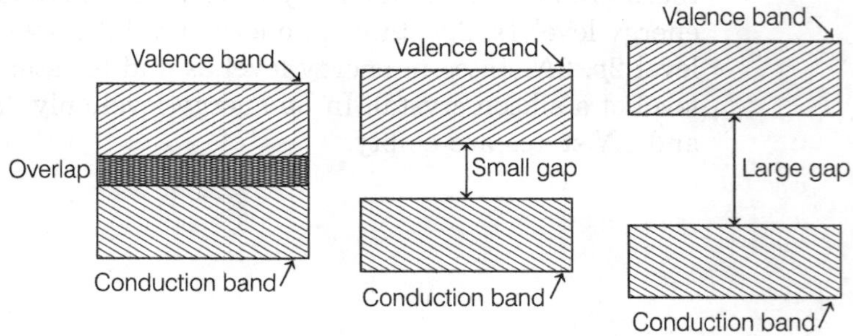

Conductor **Semiconductor** **Insulator**

Conductors	Insulators	Semiconductors
The valence and conduction bands overlap on each other.	The valence band is completely filled with electrons and the conduction band is empty.	Semiconductors are the materials in which the forbidden energy gap between the valence band and the conduction band is very small.
There is no forbidden energy gap.	Both the bands are separated by a forbidden energy gap of about 6-7 eV.	Both the bands are separated by a forbidden energy gap less than 3 eV.
As there is a large number of conduction electrons, so conductors or metals are good conductors of electricity.	Insulator is a bad conductor of electricity.	At 0 K, the electrons in the valence band do not have sufficient energy to jump to the conduction band and hence, semiconductor behaves as an insulator at 0K. But at room temperature, some of the valence electrons have sufficient thermal energy to jump to the conduction band.Hence, semiconductor may conduct at room temperature.
Resistivity increases with increase in temperature.	Resistivity decreases very slowly with increase in temperature.	Resistivity decreases with increase in temperature.

Intrinsic and Extrinsic Semiconductors

In semiconductors, the conduction band and the valence band are separated by a relatively small energy gap. For silicon, this gap is 1.1 eV and for germanium, it is 0.7 eV.

Semiconductors are divided in two groups

Intrinsic Semiconductors

A pure (free from impurity) semiconductor which has a valency 4 is called an intrinsic semiconductor. Pure germanium, silicon or carbon in their natural state are intrinsic semiconductors.

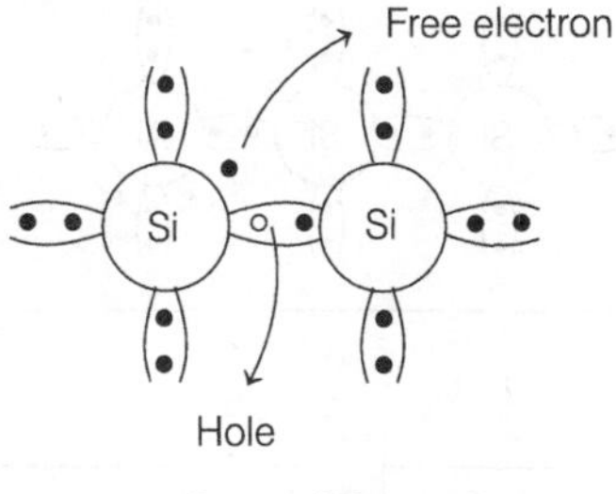

Number of holes = number of free electrons

Extrinsic Semiconductors

(i) ***p*-type semiconductors** A trivalent (for example, boron, aluminium, gallium or indium) is added to a germanium or silicon.

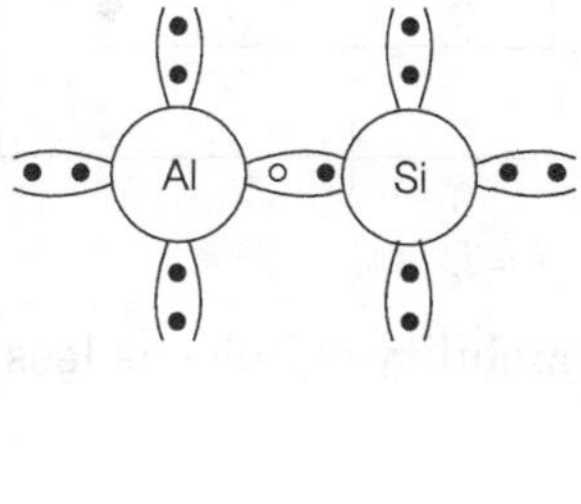

(a) Holes are the majority charge carriers.

$$n_h >> n_e$$

(b) p-type semiconductor is electrically neutral.

(c) p-type semiconductor can be shown as

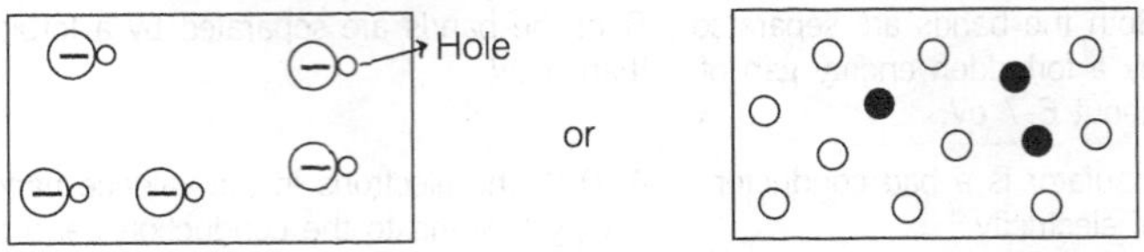

(ii) ***n*-type semiconductors** A pentavalent impurity atom (antimony, phosphorus or arsenic) is added to a Ge or Si crystal.

(a) Electrons are the majority charge carriers.

$$n_e >> n_h$$

(b) n-type semiconductor is also electrically neutral.

(c) n-type semiconductor can be shown as

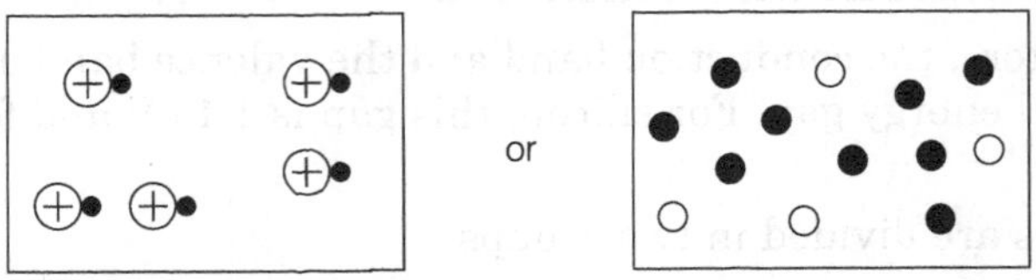

Electrical Conduction through Semiconductors

Electrical conduction through semiconductors can be understood by figure given below.

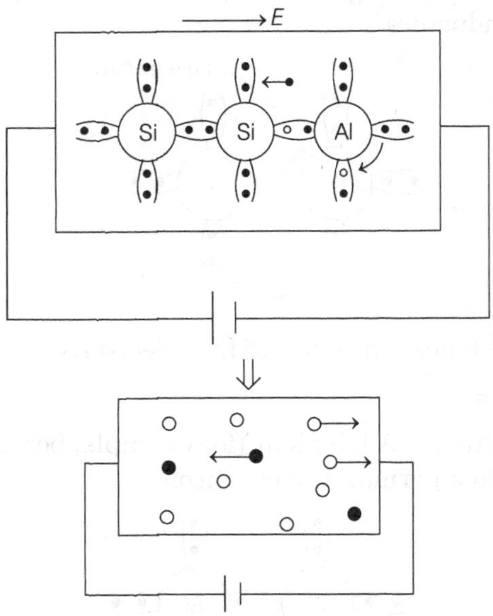

Here, $i = i_e + i_h$

But it should be noted that mobility of holes is less than the mobility of electrons.

p-n Junction Diode

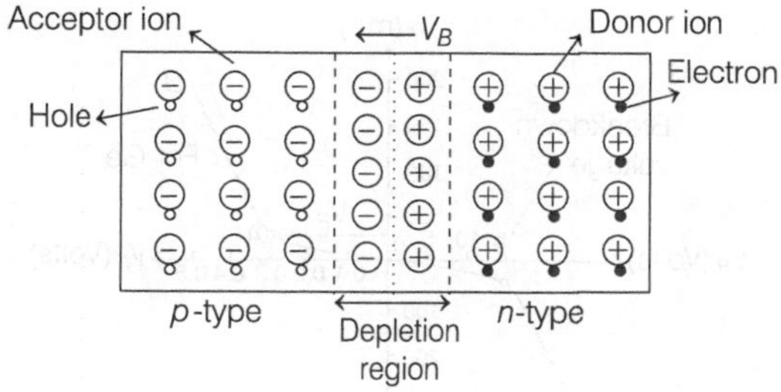

- In a *p-n* junction diode, holes are majority charge carriers on *p*-side and electrons on *n*-side. Holes, thus diffuse to *n*-side and electrons to *p*-side.
- This diffusion causes an excess positive charge in the *n* region and an excess negative charge in the *p* region near the junction. This double layer of charge creates an electric field which exerts a force on the electrons and holes, against their diffusion.
- In the equilibrium position, there is a barrier, for charge motion with the *n*-side at a higher potential than the *p*-side, it is called depletion region. There is a barrier V_B called potential barrier.
- The symbol of *p-n* junction diode is shown below

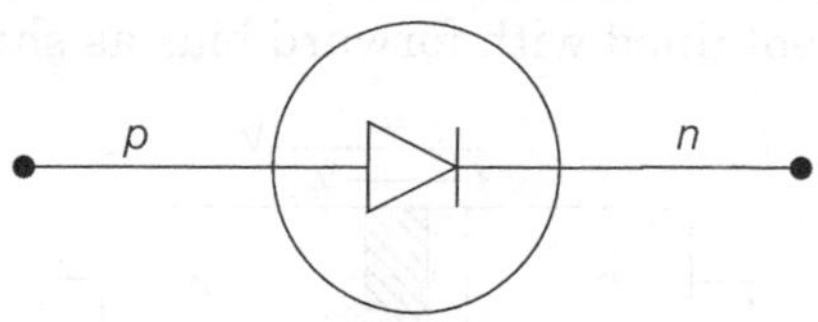

Diffusion Current and Drift Current

Because of concentration difference, holes try to diffuse from the *p*-side to the *n*-side at the *p-n* junction. This diffusion give rise to a current from *p*-side to *n*-side called diffusion current. Because of thermal collisions, electron-hole pair are created at every part of a diode.

However, if an electron-hole pair is created in the depletion region, the electron is pushed by the electric field towards the *n*-side and the hole towards the *p*-side. This gives rise to a current from *n*-side to *p*-side called the drift current.

$$I_{df} \longrightarrow \text{from } p\text{-side to } n\text{-side}$$

$$I_{dr} \longrightarrow \text{from } n\text{-side to } p\text{-side}$$

When diode is unbiased $I_{df} = I_{dr}$ or $I_{\text{net}} = 0$.

When diode is forward biased $I_{df} > I_{dr}$ or I_{net} is from *p*-side to *n*-side.

When diode is reverse biased $I_{dr} > I_{df}$ or I_{net} is from *n*-side to *p*-side.

V-*I* Characteristics of *p*-*n* Junction Diode

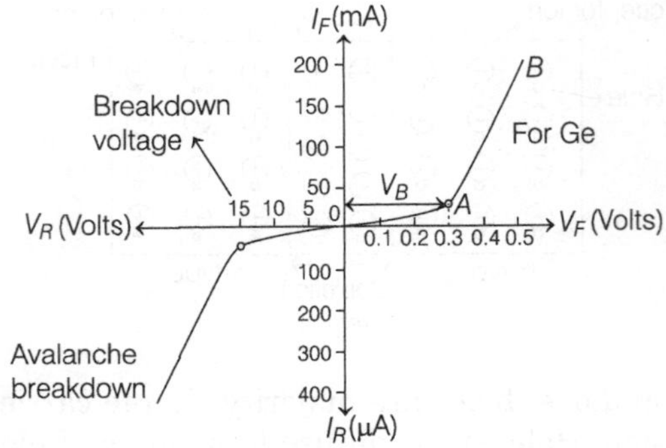

Forward Bias

In forward biased condition, *p*-type of the *p*-*n* junction is connected to the positive terminal and *n*-type is connected to the negative terminal of the external voltage.

This results in reduced potential barrier.

At some forward voltage, i.e 0.7 V for Si and 0.3 V for Ge, the potential barrier is almost eliminated and the current starts flowing in the circuit.

From this instant, the current increases with the increase in forward voltage. Hence, a curve *OB* is obtained with forward bias as shown in figure above.

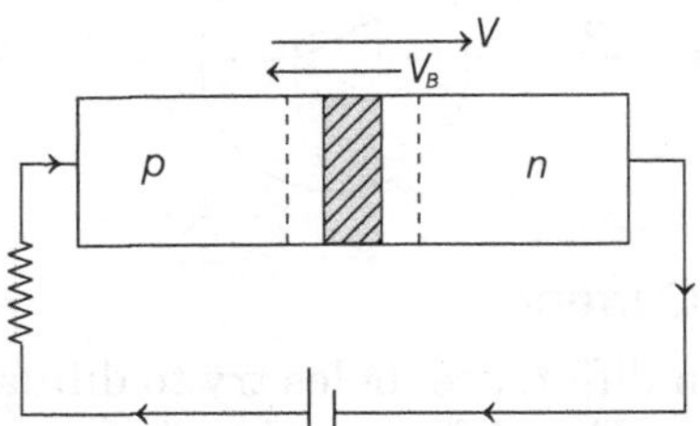

Form the forward characteristics, it can be noted that at first, i.e. region *OA*, the current increases very slowly.

However, once the external voltage exceeds the potential barrier voltage, the potential barrier is eliminated and the *p*-*n* junction behaves as an ordinary conductor. Hence, the curve *AB* rises very sharply with the increase in external voltage.

Reverse Bias

In reverse bias condition, the *p*-type of the *p*-*n* junction is connected to the negative terminal and *n*-type is connected to the positive terminal of the external voltage.

This results in increased potential barrier at the junction.

Hence, the junction resistance becomes very high and as a result practically no current flows through the circuit.

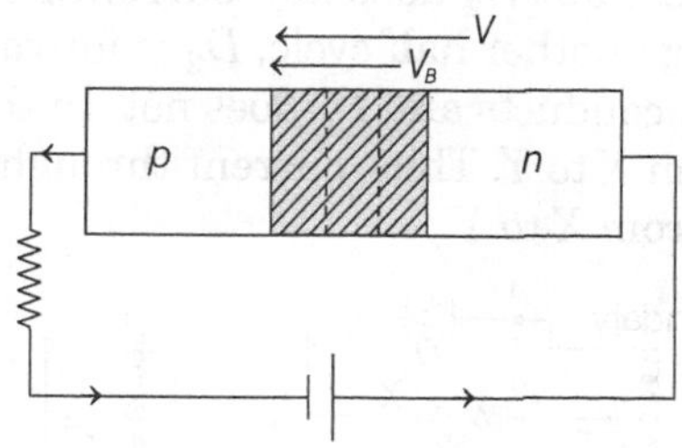

However, a very small constant current of the order of μA, flows through the circuit due to minority charge carriers on both sides (electrons on p-side and holes on n-side). This is known as reverse saturation current(I_s).

The reverse bias applied to the p-n junction acts as forward bias to these minority carriers and hence, small current flows in the reverse direction.

If the applied reverse voltage is increased continuously, the kinetic energy of the minority carriers may become high enough to knock out electrons from the semiconductor atom.

At this stage, breakdown of the junction may occur. This is characterised by a sudden increase of reverse current and a sudden fall of the resistance of barrier region. This phenomenon is called avalanche breakdown and the voltage beyond which current suddenly increases is called breakdown voltage.

At breakdown voltage, the current through diode shoots rapidly. Even for a small change in applied voltage, there is a high increase in net current through the diode.

Junction Diode as a Rectifier

A rectifier is a device which converts an alternating current (or voltage) into a direct (or unidirectional) current (or voltage). It offers a low resistance for the current to flow when it is forward biased, but a very high resistance when reverse biased. Thus, it allows current through it only in one direction and acts as a rectifier.

(i) ***p-n* junction diode as half-wave rectifier**

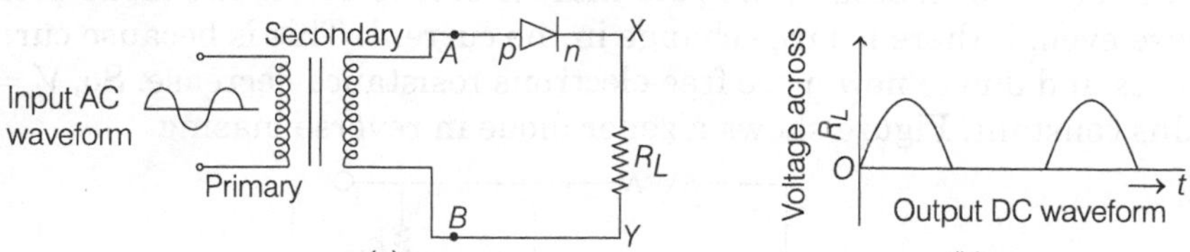

(a) (b)

When the voltage at A is positive, the diode is forward biased and it conducts and when the voltage at A is negative, the diode is reverse biased and does not conduct. Thus, current through R_L flow only in one direction from X to Y though only in half cycle.

(ii) ***p-n* junction diode as full-wave rectifier**

During one-half cycle, D_1 is forward biased and D_2 is reverse biased. Therefore, D_1 conducts but D_2 does not, current flows from X to Y through load resistance R_L. During another half cycle, D_2 is forward biased and D_1 reverse biased. Therefore, D_2 conducts and D_1 does not. In this half cycle also current through R_L flows from X to Y. Thus, current through R_L in both the half cycles is in one direction, i.e. from X to Y.

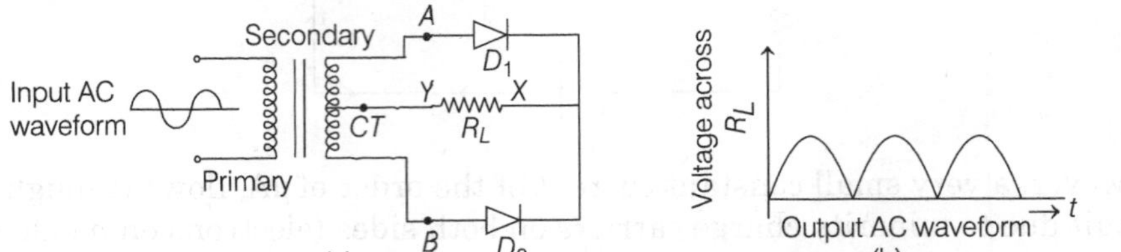

(a) (b)

Bridge Rectifier

Current through R_L always flows in one direction from X to Y.

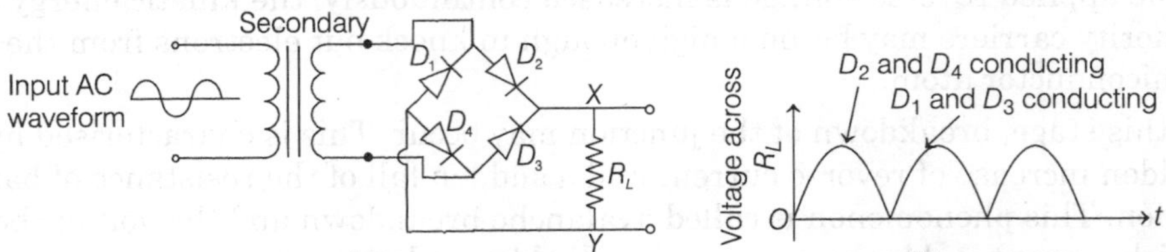

(a) (b)

Note *Even after rectification, ripples are present in the output which can be removed upto great extent by a filter circuit. A filter circuit consists of a capacitor.*

Zener Diode

A diode meant to operate under reverse bias in the breakdown region is called an avalanche diode or a zener diode. Such diode is used as a voltage regulator. The symbol of zener diode is shown in figure

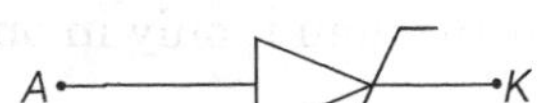

Once the breakdown occurs, the potential difference across the diode does not increase even, if there is large change in the current. This is because current increases and due to new more free electrons resistance decrease. So, $V = IR$ remains constant. Figure shows a zener diode in reverse biasing.

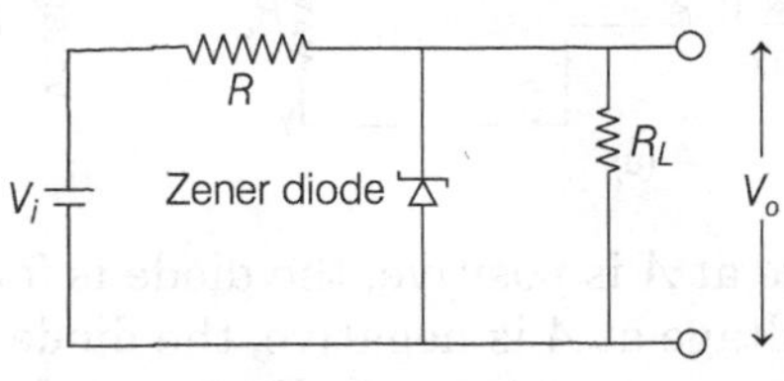

An input voltage V_i is connected to the zener diode through a series resistance R such that the zener diode is reverse biased.

If the input voltage increases, the current through R and zener diode also increases. This increases the voltage drop across R without any change in the voltage across the zener diode. Similarly, if the input voltage decreases the current through R and zener diode also decreases. The voltage drop across R decreases without any change in the voltage across the zener diode.

Thus any increase/decrease in the input voltage results in increase/decrease of the voltage drop across R without any change in voltage across the zener diode (and hence across load resistance R_L). Thus, the zener diode acts as a voltage regulator.

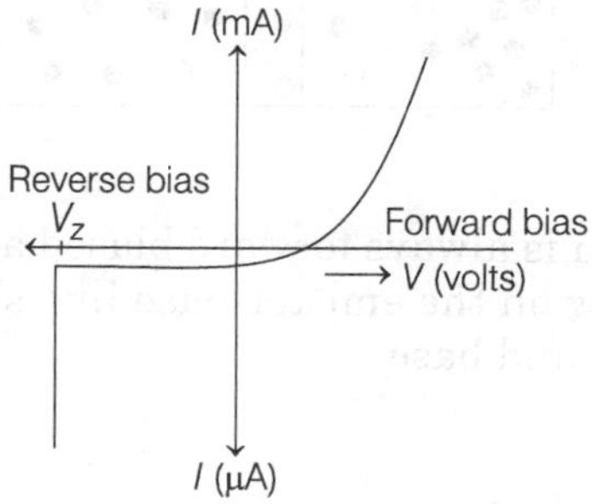

Junction Transistors

A transistor is a semiconductor device used to amplify or switch electronic signal and electrical power. There are two types of transistors

(i) *p-n-p* Transistor

When a thin layer of *n*-type semiconductor sandwiched between two *p*-type semiconductors, then it is known as *p-n-p* transistor.

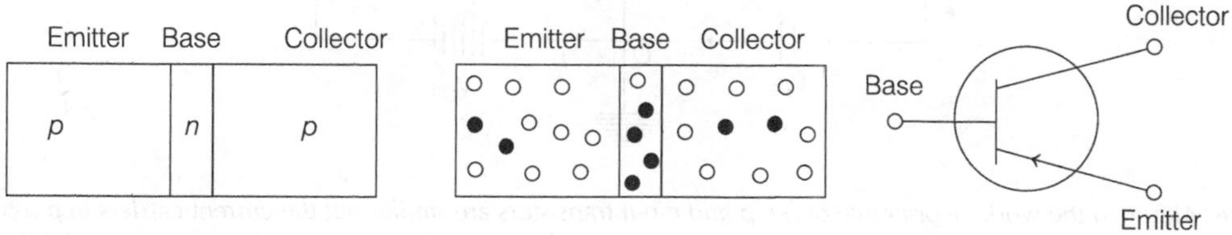

Working of a *p-n-p* Transistor

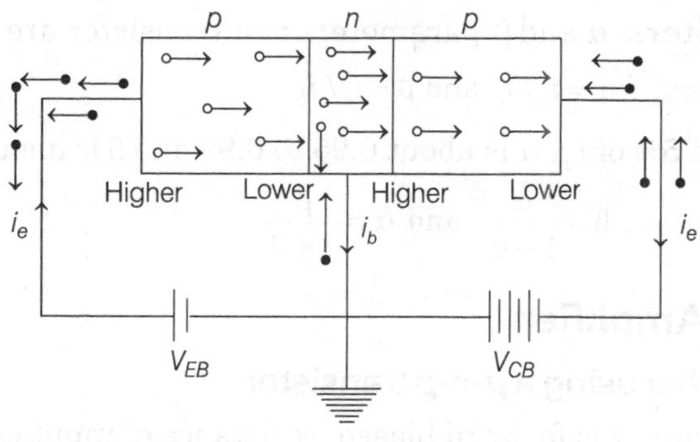

$$i_e = i_c + i_b$$

Note *i_b is only about 2% of i_e or roughly around 2% of holes coming from emitter to base combine with the electrons. Rest 98% move to collector.*

(ii) *n-p-n* Transistor

When a thin layer of p-type semiconductor sandwiched between two n-type semiconductors, then it is called n-p-n transistor.

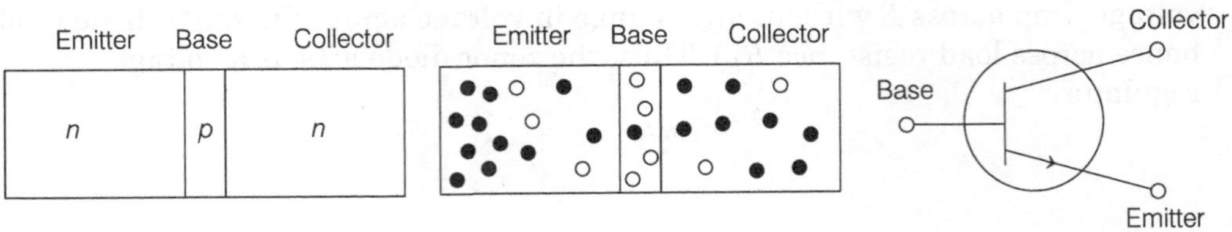

The emitter-base junction is always forward biased and collector-base junction is reverse biased. The arrow on the emitter-base line shows the direction of current between emitter and base.

Working of *n-p-n* Transistor

$$i_e = i_b + i_c$$

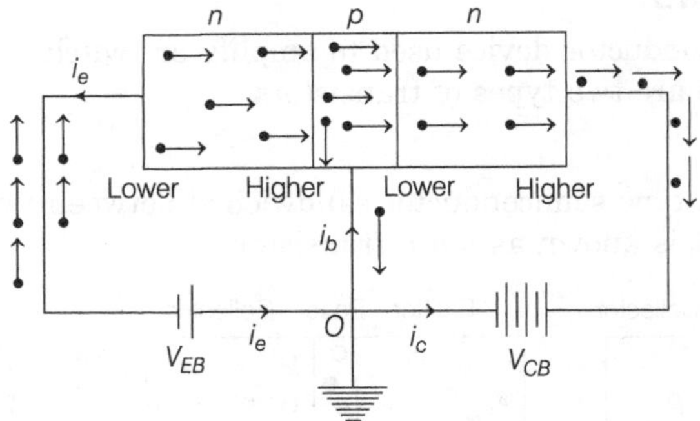

Note *Although the working principle of p-n-p and n-p-n transistors are similar but the current carriers in p-n-p transistor are mainly holes whereas in n-p-n transistors the current carriers are mainly electrons. Mobility of electrons are however more than the mobility of holes, therefore n-p-n transistors are used in high frequency and computer circuits where the carriers are required to respond very quickly to signals.*

α and β-parameters α and β-parameters of a transistor are defined as

$$\alpha = i_c / i_e \text{ and } \beta = i_c / i_b$$

As i_b is about 1 to 5% of i_e, α is about 0.95 to 0.99 and β is about 20 to 100.

$$\beta = \frac{\alpha}{1-\alpha} \text{ and } \alpha = \frac{\beta}{1+\beta}$$

Transistor As An Amplifier

Common base amplifier using a *p-n-p* transistor

Since, the input circuit is forward biased, resistance of input circuit is small. Similarly, output circuit is reverse biased, hence resistance of output circuit is high.

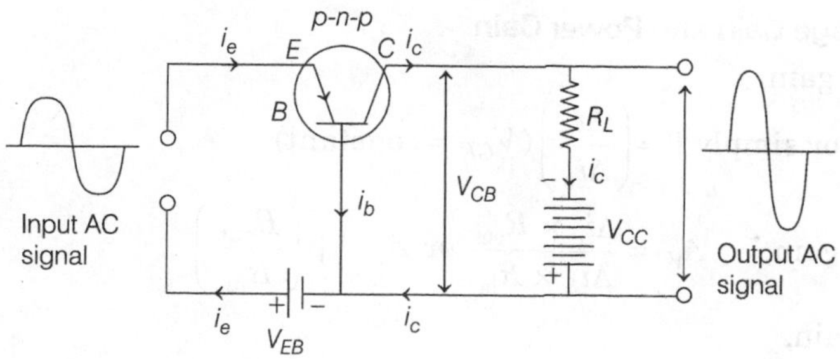

$$V_{CB} = V_{CC} - i_c R_L$$

Due to fluctuations in V_{EB}, the emitter current i_e also fluctuates which in turn fluctuates i_c. In accordance with the above equation, there are fluctuations in V_{CB}.

Current gain, Voltage gain and Power gain

- Current gain, α_{AC} or simply $\alpha = \dfrac{\Delta i_c}{\Delta i_e}$ (V_{CB} = constant)
- Voltage gain, $A_V = \dfrac{\Delta i_c \times R_{out}}{\Delta i_e \times R_{in}}$ but $\dfrac{\Delta i_c}{\Delta i_e} = \alpha$

$$\therefore \quad A_V = \frac{\alpha R_{out}}{R_{in}}$$

Since, $R_{out} >> R_{in}$, A_V is quite high, although α is slightly less than 1.

- Power gain

$$P = Vi$$

Therefore, power gain = current gain × voltage gain

$$\text{Power gain} = \alpha^2 \cdot \frac{R_{out}}{R_{in}}$$

The output voltage signal is in phase with the input voltage signal.

Common emitter amplifier using a *p-n-p* transistor

Since, the base-emitter circuit is forward biased, input resistance is low.

Similarly, collector-emitter circuit is reverse biased, therefore output resistance is high.

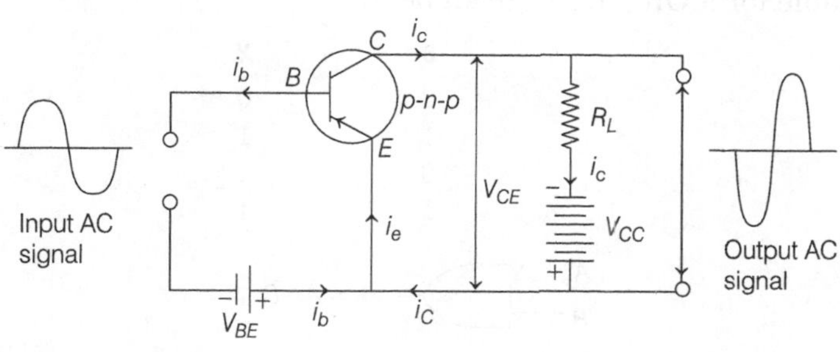

$$V_{CE} = V_{CC} - i_c R_L$$

Current Gain, Voltage Gain and Power Gain

- Current gain,

$$\beta_{AC} \text{ or simply } \beta = \left(\frac{\Delta i_c}{\Delta i_b}\right) (V_{CE} = \text{constant})$$

- Voltage gain, $$A_V = \frac{\Delta i_c \times R_{out}}{\Delta i_b \times R_{in}} \text{ or } A_V = \beta\left(\frac{R_{out}}{R_{in}}\right)$$

- Power gain,

$$P = Vi$$

Therefore, power gain = current gain × voltage gain

or $$\text{Power gain} = \beta^2 \left(\frac{R_{out}}{R_{in}}\right)$$

The value of current gain β is from 15 to 50 which is much greater than α.

Note *(i) The voltage gain in common emitter amplifier is larger compared to that in common base amplifier.*
(ii) The power gain in common emitter amplifier is extremely large compared to that in common base amplifier.
(iii) The output voltage signal is 180° out of phase with the input voltage signal in the common emitter amplifier.

Transconductance (g_m)

There is one more term called transconductance (g_m) in common emitter mode.

$$g_m = \left(\frac{\Delta i_c}{\Delta V_{BE}}\right) (V_{CE} = \text{constant})$$

The unit of g_m is Ω^{-1} or siemen (S).

$$g_m = \frac{\beta}{R_{in}}$$

Logic gates

A digital circuit which allows a signal to pass through it, only when a few logical relations are satisfied is called a logic gate.These logic gates are as follows

OR gate

The truth table for a OR gate is given below

A	B	X
0	0	0
0	1	1
1	0	1
1	1	1

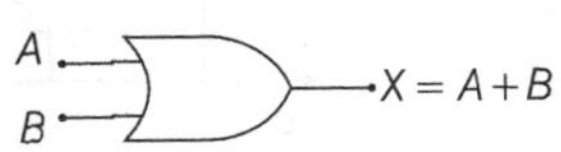

It is written as $X = A + B$

Figure shows construction of an OR gate using two diodes D_1 and D_2

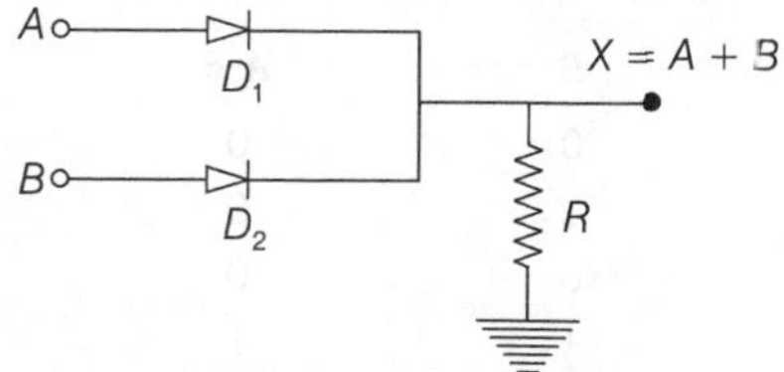

AND gate

The truth table for a AND gate is given below

A	B	Y
0	0	0
0	1	0
1	0	0
1	1	1

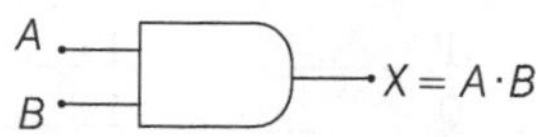

It is written as $X = A \cdot B$

Figure shows construction for an AND gate using two diodes D_1 and D_2.

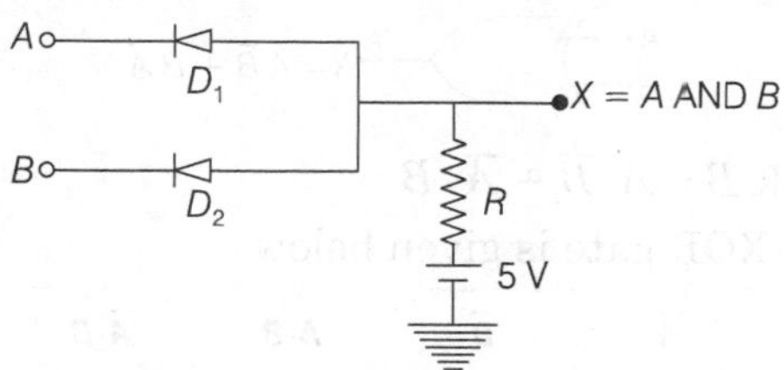

NOT gate

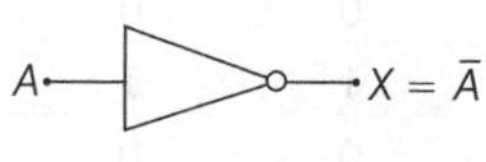

It is written as, $X = \overline{A}$

The truth table for a NOT gate is given below

A	X
0	1
1	0

Note *A NOT gate cannot be constructed with diodes. Transistor is used for realisation of a NOT gate.*

NAND gate

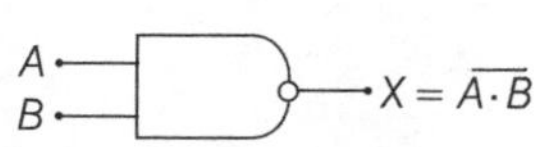

It is written as $X = \overline{A \cdot B}$ or $X = \overline{AB}$

The truth table for a NAND gate is given below

A	B	$A \cdot B$	$X = \overline{A \cdot B}$
0	0	0	1
0	1	0	1
1	0	0	1
1	1	1	0

NOR gate

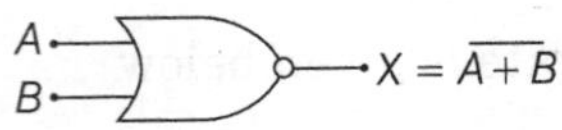

It is written as $X = \overline{A + B}$

The truth table for a NOR gate is given below

A	B	$A + B$	$X = \overline{A + B}$
0	0	0	1
0	1	1	0
1	0	1	0
1	1	1	0

XOR gate

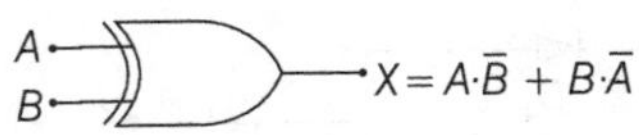

It is written as $A \text{XOR } B = A \cdot \overline{B} + \overline{A} \cdot B$

The truth table for a XOR gate is given below.

A	B	$\overline{A}$	$\overline{B}$	$A \cdot \overline{B}$	$\overline{A} \cdot B$	$A = A \cdot \overline{B} + \overline{A} \cdot B$
0	0	1	1	0	0	0
0	1	1	0	0	1	1
1	0	0	1	1	0	1
1	1	0	0	0	0	0